T0192421

Lecture Notes in Computer Science 14503

Founding Editors

Gerhard Goos
Juris Hartmanis

Editorial Board Members

The series Lecture Notes in Computer Science (LNCS), including its subseries Lecture Notes in Artificial Intelligence (LNAI) and Lecture Notes in Bioinformatics (LNBI), has established itself as a medium for the publication of new developments in computer science and information technology research, teaching, and education.

LNCS enjoys close cooperation with the computer science R & D community, the series counts many renowned academics among its volume editors and paper authors, and collaborates with prestigious societies. Its mission is to serve this international community by providing an invaluable service, mainly focused on the publication of conference and workshop proceedings and postproceedings. LNCS commenced publication in 1973.

Hai Jin · Zhiwen Yu · Chen Yu · Xiaokang Zhou ·
Zeguang Lu · Xianhua Song
Editors

Green, Pervasive, and Cloud Computing

18th International Conference, GPC 2023
Harbin, China, September 22–24, 2023
Proceedings, Part I

 Springer

Editors
Hai Jin
Huazhong University of Science
and Technology
Wuhan, China

Chen Yu
Huazhong University of Science
and Technology
Wuhan, China

Zeguang Lu
National Academy of Guo Ding Institute
of Data Science
Beijing, China

Zhiwen Yu
Harbin Engineering University
Harbin, China

Xiaokang Zhou
Shiga University
Shiga, Japan

Xianhua Song
Harbin University of Science and Technology
Harbin, China

ISSN 0302-9743   ISSN 1611-3349 (electronic)
Lecture Notes in Computer Science
ISBN 978-981-99-9892-0   ISBN 978-981-99-9893-7 (eBook)
https://doi.org/10.1007/978-981-99-9893-7

This Springer imprint is published by the registered company Springer Nature Singapore Pte Ltd.
The registered company address is: 152 Beach Road, #21-01/04 Gateway East, Singapore 189721, Singapore

Paper in this product is recyclable.

Preface

As the program chairs of the 18th International Conference on Green, Pervasive, and Cloud Computing (GPC 2023), it is our great pleasure to welcome you to the proceedings of the conference, which was held in Harbin, China, 22–24 September 2023, hosted by Huazhong University of Science and Technology, Harbin Engineering University, Harbin Institute of Technology, Northeast Forestry University, Harbin University of Science and Technology, Heilongjiang Computer Federation and National Academy of Guo Ding Institute of Data Sciences. The goal of this conference was to provide a forum for computer scientists, engineers and educators.

This conference attracted 111 paper submissions. After the hard work of the Program Committee, 38 papers were accepted to be presented at the conference, with an acceptance rate of 34.23%. There were at least 3 reviewers for each article, and each reviewer reviewed no more than 5 articles. The major topic of GPC 2023 was Green, Pervasive, and Cloud Computing. The accepted papers cover a wide range of areas related to Edge Intelligence and Mobile Sensing and Computing, Cyber-Physical-Social Systems and Industrial Digitalization and Applications.

We would like to thank all the program committee members of both conferences, for their hard work in completing the review tasks. Their collective efforts made it possible to attain quality reviews for all the submissions within a few weeks. Their expertise in diverse research areas helped us to create an exciting program for the conference. Their comments and advice helped the authors to improve the quality of their papers and gain deeper insights.

We thank Xuemin Lin, Nabil Abdennadher and Laurence T. Yang, whose professional assistance was invaluable in the production of the proceedings. Great thanks should also go to the authors and participants for their tremendous support in making the conference a success. Besides the technical program, this year GPC offered different experiences to the participants. We hope you enjoyed the conference.

September 2023

Zhiwen Yu
Christian Becker
Chen Yu
Xiaokang Zhou
Chengtao Cai

Organization

Honorary Chair

Rajkumar Buyya   University of Melbourne, Australia

General Chairs

Zhiwen Yu   Harbin Engineering University, China
Christian Becker   University of Stuttgart, Germany

Program Chairs

Chen Yu   Huazhong University of Science and Technology, China
Xiaokang Zhou   Shiga University, Japan
Chengtao Cai   Harbin Engineering University, China

Organization Chairs

Wei Wang   Harbin Engineering University, China
Haiwei Pan   Harbin Engineering University, China
Hongtao Song   Harbin Engineering University, China
Yuhua Wang   Harbin Engineering University, China
Jiquan Ma   Heilongjiang University, China
Xingyu Feng   NSFOCUS, China

Publication Chairs

Weixi Gu   University of California, Berkeley, USA
Xianhua Song   Harbin University of Science and Technology, China
Dan Lu   Harbin Engineering University, China

Sponsorship Chair

Bingyang Li   Harbin Engineering University, China

Session Chairs

Chen Yu   Huazhong University of Science and Technology, China

Xianwei Lv   Huazhong University of Science and Technology, China

Web Chair

Wang Chen   Huazhong University of Science and Technology, China

Registration/Financial Chair

Fa Yue   National Academy of Guo Ding Institute of Data Science, China

Steering Committee

Hai Jin (Chair)   Huazhong University of Science and Technology, China

Nabil Abdennadher (Executive Member)   University of Applied Sciences, Switzerland

Christophe Cerin   University of Paris Xlll, France

Sajal K. Das   Missouri University of Science and Technology, USA

Jean-Luc Gaudiot   University of California-Irvine, USA

Kuan-Ching Li   Providence University, Taiwan

Cho-Li Wang   University of Hong Kong, China

Chao-Tung Yang   Tunghai University, Taiwan

Laurence T. Yang   St. Francis Xavier University, Canada/Hainan University, China

Zhiwen Yu   Harbin Engineering University, China

Program Committee Members

Alex Vieira	Federal University of Juiz de Fora, Brazil
Alfredo Navarra	Università degli Studi di Perugia, Italy
Andrea De Salve	Institute of Applied Sciences and Intelligent Systems, Italy
Arcangelo Castiglione	University of Salerno, Italy
Bo Wu	Tokyo University of Technology, Japan
Carson Leung	University of Manitoba, Canada
Feng Tian	Chinese Academy of Sciences, China
Florin Pop	University Politehnica of Bucharest, Romania
Jianquan Ouyang	Xiangtan University, China
Ke Yan	National University of Singapore, Singapore
Kuo-chan Huang	National Taichung University of Education, Taiwan
Liang Wang	Northwestern Polytechnical University, China
Marek Ogiela	AGH University of Science and Technology, Poland
Mario Donato Marino	Leeds Beckett University, England
Meili Wang	Northwest A&F University, China
Pan Wang	Nanjing University of Posts and Telecommunications, China
Shingo Yamaguchi	Yamaguchi University, Japan
Su Yang	Fudan University, China
Tianzhang Xing	Northwest University, China
Wasim Ahmad	University of Glasgow, UK
Weixi Gu	China Academy of Industrial Internet, China
Xiaofeng Chen	Xidian University, China
Xiaokang Zhou	Shiga University, Japan
Xiaotong Wu	Nanjing Normal University, China
Xin Liu	China University of Petroleum, China
Xujun Ma	Télécom SudParis, France
Yang Gu	Chinese Academy of Sciences, China
Yanmin Zhu	Shanghai Jiao Tong University, China
Yaokai Feng	Kyushu University, Japan
Yirui Wu	Hohai University, China
Yuan Rao	Anhui Agricultural University, China
Yuezhi Zhou	Tsinghua University, China
Yufeng Wang	Nanjing University of Posts and Telecommunications, China
Yunlong Zhao	Nanjing University of Aeronautics and Astronautics, China

Zengwei Zheng   Hangzhou City University, China
Zeyar Aung   Khalifa University, United Arab Emirates
Zhijie Wang   Hong Kong Polytechnic University, China

Contents – Part I

Contents – Part II

Cyber-Physical-Social Systems

Pervasive and Green Computing

Wireless and Ubiquitous Networking

Industrial Digitization and Applications

UEBCS: Software Development Technology Based on Component Selection

Yingnan Zhao[1], Xuezhao Qi[1], Jian Li[2(✉)], and Dan Lu[1]

[1] College of Computer Science and Technology, Harbin Engineering University,
Harbin 150009, China
{zhaoyingnan,ludan}@hrbeu.edu.cn
[2] Harbin Institute of Technology Software Engineering Co., Ltd., Harbin, China
13009860167@163.com

Abstract. Current software development has moved away from the traditional manual workshop model and emphasizes improving software product quality. To address the issue of repetitive work, software reuse techniques can be adopted to continually enhance the quality and efficiency of software development. Software reuse primarily involves reutilizing existing software knowledge during the software development process, effectively reducing maintenance costs incurred during development and controlling the overall software development expenses. Software components are an effective form of supporting software product reuse and serve as the core technology for enabling software reuse. Component-based software engineering techniques emphasize the use of reusable software "components" to design and construct programs, aiming to assemble these components within a software architecture to achieve software reuse and improve the quality and productivity of software products. However, selecting the most suitable components from the results of component retrieval requires understanding the different usages of each component in the retrieval results. The existing methods suffer from excessive reliance on manual approaches and errors caused by inter-component relationships. Therefore, this paper proposes a component selection technique called UEBCS (Usage Example-Based Component Selection). This technique leverages steps such as clustering analysis and hierarchical classification to achieve optimal component selection. UEBCS has shown excellent results in terms of both efficiency and accuracy in selecting components. This method provides technical support for software developers in the software development process and holds significant practical significance for enhancing software development quality and efficiency, as well as promoting the deepening development of the software industry.

Keywords: Software Engineering · Software Reuse · Software Component

1   Introduction

With the transition from procedural to object-oriented programming, software has become increasingly large and complex. However, contrary to expectations, the "software crisis" has not disappeared and has instead become more challenging due to the existence of legacy systems. Since the inception of the software industry, there has been a shift from single-task work to engineering operations, and the accompanying "software crisis" has become the most critical issue in software engineering, with legacy systems being the most difficult problem to solve. Software reuse technology is an important support for addressing this key problem and is the most effective solution to the software crisis in software engineering, with component technology being the core technology of software reuse and the foundation of advanced software engineering [1]. Software components are independent, pluggable, and reusable software units with specific functionality and interfaces [2]. Components form the foundation of advanced software engineering. The rise of component technology stems from the engineering concept of software reuse, which originated from the need to handle legacy system problems [3]. Components offer a broader perspective compared to objects and procedures; they provide a higher-level approach to software development and are better suited for engineering thinking in software engineering. Adopting component-based software reuse technology helps alleviate the software crisis and holds profound implications for the future development of software engineering [4]. Component technology is the core technology of software reuse, and components can range from small units such as functions, classes, and objects to complete software systems [5].

To address this key problem, this paper proposes a method called UEBCS for selecting components based on their different usage examples, with the goal of achieving software reuse. In this method, to obtain usage examples of components, relevant code is first gathered from the Internet as candidate examples. Then, cluster analysis is performed on the candidate examples. Subsequently, one representative example is selected from each cluster. Finally, based on a grading strategy, the representative examples are categorized into different levels. These representative examples provide developers with insights into the various usages of components, thereby promoting software reuse. We conducted a simulation test of our method on an open-source project and achieved excellent results in terms of both efficiency and accuracy in component selection.

The structure of this paper is as follows: Section 2 introduces related work about software reuse and software components. Section 3 provides a detailed description of the specific steps of the UEBCS method. Section 4 concludes the paper by summarizing the proposed method.

2   Related Work

Software reuse techniques have been applied in practical software development processes. Open-source component libraries are a collection of softwarecomponents shared and maintained by developers. Developers can utilize these components to build their own applications, thereby saving development time and cost. Common open-source component libraries include React, Angular, Vue, and others. Third-party libraries and frameworks are software components built and maintained by other developers, offering various functionalities and tools. Developers can leverage these libraries and frameworks to expedite the development process and minimize redundant efforts. Common third-party libraries and frameworks include jQuery, Bootstrap, Django, and others.

Research on software component selection techniques has been ongoing for many years, and researchers have proposed various methods to support software reuse. Comella-Dorda et al. introduced a business-based component selection method [6]. However, relying solely on business-provided information often leads to the selection of components that do not meet the actual functional requirements. Ballurio et al. proposed a manual evaluation method for selecting components [7]. However, an excessive reliance on manual approaches can introduce inaccuracies due to subjective recommendations and result in lower efficiency. Although Mancebo et al. presented an approach that employs end-to-end automation to address the efficiency issue, they did not consider the interdependencies among software components, which can lead to errors during the reuse process of the selected software components [8].

To address the limitations of the aforementioned methods, we propose an approach called UEBCS. UEBCS is a fully automated method that takes into account the interrelationships among software components. By integrating clustering algorithms, it resolves the issues of excessive reliance on manual approaches and component errors caused by inter-component relationships during the reuse process.

2.1 SoftWare Reuse

As software systems continue to grow in size and software development processes become more complex, software reuse technology becomes increasingly important. Software reuse refers to the practice of reusing existing software assets, components, modules, or designs during the software development process to reduce effort, improve development efficiency, and enhance quality [9]. It is a crucial practice in the field of software engineering, aiming to address the "software crisis," reduce development costs, accelerate time to market, and improve the maintainability and scalability of software systems [10].

The benefits of software reuse include [11]:

1) Improved development efficiency: By reusing existing software assets, repetitive work can be reduced, saving development time and costs.
2) Enhanced software quality: Reusing tested and validated components can improve the reliability and stability of the system, reducing errors and defects.
3) Accelerated time to market: By leveraging existing software assets, development cycles can be shortened, allowing for rapid product releases.
4) Reduced maintenance costs: Reused software components often have a higher level of maintainability, as they have been used and tested multiple times. This can lower the costs and effort involved in system maintenance.

5) Facilitates standardization and modularity: Software reuse encourages the modularization of functionality, making systems easier to understand, extend, and maintain. It also promotes the establishment and adherence to industry standards.

Software reuse can be achieved through various methods, including:

1) Component library and asset management: Establishing a component library or asset repository to store and manage reusable software components and assets, making it easier for developers to search for and use them.
2) Interface definition and standardization: Defining clear interfaces and standards to ensure interoperability and consistency of interactions between software components.
3) Domain engineering and model-driven approaches: Transforming domain knowledge and models into reusable software components through domain engineering and model-driven approaches.
4) Utilizing open-source software and third-party libraries: Leveraging open-source software and third-party libraries as pre-validated and tested reusable components.
5) Componentization and modularization approaches: Dividing the software system into independent components or modules for reuse and composition in different systems.

These methods provide ways to effectively utilize and manage reusable software components, leading to increased development efficiency, improved quality, and reduced development costs.

Software reuse also faces several challenges, including component selection, component adaptation, and component integration [12]. Therefore, organizing and managing reusable components effectively is a crucial factor for successfully implementing software reuse.

In conclusion, software reuse is an effective software development practice that involves leveraging existing software assets to improve development efficiency, reduce costs, and enhance the quality and maintainability of software systems. It plays a crucial role in modern software engineering, providing developers with a reliable approach to building reliable and efficient software systems.

2.2   Software Component

A component refers to a unit of software that is semantically complete, syntactically correct, and has reusable value. It is a discernible constituent in the process of reuse. Components possess relative independence, functionality, and interchangeability. They are not dependent on a specific system and are not specific or proprietary but have practical functional significance and can be directly replaced by identical components. A component is a relatively independent and transmissible collection of operations, a replaceable software unit. Components encompass various design roles, and they can be part of a larger unit or composed of multiple components. Within a component, there are numerous attributes, including ports, types, constraints, semantics, and evolution. A component is not a complete application; it is only when multiple components are interconnected that they form a complete application. Software components are independent and reusable modules in a software system, representing specific functional units. Components can be developed, tested, deployed, and maintained independently and can be integrated and interact with other components.

Components are the most fundamental elements that make up a software system, and Fig. 1 shows the component modeling diagram. The "interface-port" represents the interface-style port of the component, "type" denotes the type of the component, "language" describes the semantic representation of the component, "restraints" specify the constraints and limitations of the component, "evolving" refers to the evolutionary form of the component, and "non-function" represents the non-functional attributes of the component.

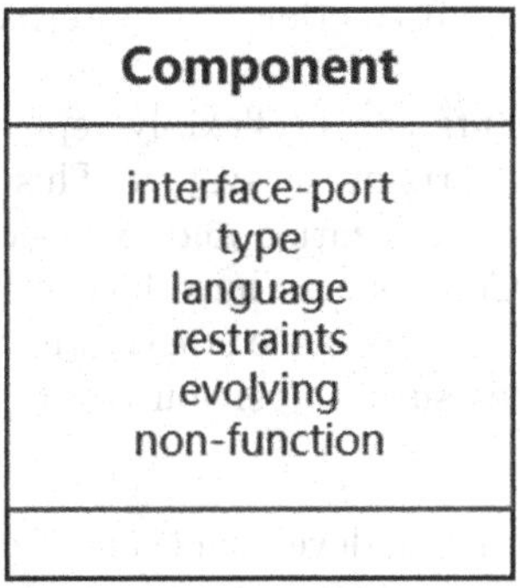

Fig. 1. Component modeling diagram.

Traditional software development techniques typically target specific engineering projects, while software component technology focuses on the common requirements of a particular industry domain. By leveraging domain engineering techniques and summarizing industry experience, domain-specific system architectures, business models, and component libraries are established to meet the industry's needs. These components possess shared characteristics within the industry domain and demonstrate strong reusability. Consequently, components serve as reusable software modules in the software development process, adhering to binary standards rather than relying on a specific high-level programming language. Component-based software technology involves accessing and implementing component functionality through interfaces in software development [13]. Software components are not dependent on a specific system but exhibit interchangeability, independence, and functionality. The widespread adoption of component-based software technology in software development has not only reduced development costs and improved efficiency but also facilitated significant enhancements in software development quality.

Software components possess the following characteristics [14]:

1) Independence: Components are independent entities that can exist and be developed and maintained separately. Component developers can design, implement, and test components independently without being affected by other components.
2) Reusability: Components are reusable and can be reused in multiple software systems or applications. By using components, developers can avoid duplicating similar functionalities, thus improving development efficiency and quality.
3) Interface Definition: Components communicate and interact with other components through explicitly defined interfaces. Interfaces define the functionalities and services exposed by the component and how other components can interact with it.
4) Encapsulation: Components encapsulate their internal implementation details, hiding the implementation specifics and internal data. This allows components to provide an abstraction layer interface, simplifying integration and usage.
5) Plug-and-Play: Components can be flexibly replaced or added to a system without impacting other parts of the system. This enables system scalability and flexibility, allowing customization and extension as needed.
6) Lifecycle Management: Components have lifecycles, including development, testing, deployment, and maintenance stages. Lifecycle management of components involves activities such as version control, documentation, testing, and troubleshooting.

By utilizing these components, developers can accelerate development speed, reduce development costs, and enhance the quality and maintainability of software systems. Component-based development also promotes modular thinking and the application of reuse principles in software engineering. In component-based software development techniques, components play a crucial role in facilitating system development and evolution.

3   Methodology

UEBCS primarily achieves the provision of code examples to help developers understand different usages of components through three steps, as illustrated in Fig. 2:

Step 1: Code Retrieval Module: This module enables the selection of candidate example code from downloaded source code on the Internet.
Step 2: Code Analysis Module: This module performs clustering of the candidate examples and selects a representative usage. The clustering is primarily based on the frequency of operation calls and comment information.
Step 3: Result Ranking: This step involves categorizing the examples into different levels and returning a list of results. The examples are ranked based on their relevance or quality.

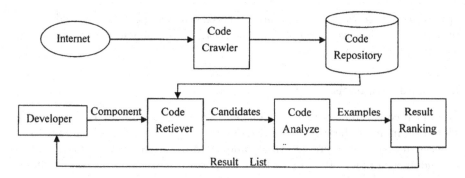

Fig. 2. UEBCS model structure.

3.1   Code Retrieval

In the UEBCS method, obtaining the code of components and the code related to the components is indeed the first step. It is true that a large amount of such information can be found on many code repository websites, making it relatively easy to obtain them. These code repository websites provide a wide range of software projects and open-source components, which allow us to browse their code repositories, documentation, and examples to access the desired code and information related to using the components. By using the search functionality on code repository websites, we can search for components and code that are relevant to our needs based on keywords, programming languages, tags, and other criteria. Obtaining code and related information from these code repository websites can help us understand the usage, functionality, and interfaces of the components. It is important to note that when using code from code repository websites, we should adhere to the corresponding licenses and legal regulations, ensuring compliance and respecting intellectual property rights. Additionally, it is crucial to carefully read and understand the documentation and examples provided with the code to ensure the proper use and integration of the components. Among these extensive collections of source code, there are certainly many examples that demonstrate different usages of components. Therefore, by leveraging the abundant resources available on the Internet, we can obtain these examples to assist developers in selecting suitable components. The advantages of using source code resources from the Internet include the following:

1) Cost-effective automatic acquisition of relevant code.
2) Access to examples showcasing component usage in various application domains.
3) Regular updates of component examples by periodically downloading source code from the Internet.

We can effectively obtain code by utilizing relevant software and storing the downloaded source code in a code repository. At any time, developers can retrieve the required component code examples from the code repository. In our approach, code retrieval is based on the class hierarchy, meaning that code examples involving the invocation of any class within a component will be considered as candidate code examples.

3.2 Code Analysis

Too much code often leads to a significant number of candidate code examples. Moreover, many of these examples demonstrate the same usage of components. It is necessary to select representative examples from the candidate set and remove redundant examples. To achieve this, we perform clustering on the candidate examples based on the usage of components. Then, we select one representative example from each cluster. By clustering, we can eliminate examples that display duplicate usages.

To identify candidate examples with different usages of components, it is essential to determine some features that can represent the usage of components. By calculating the similarity of these features, we can cluster the candidate examples.

Similarity Calculation of Call Frequency Information. If a developer wants to reuse a component, they will invoke different operations of that component. If two developers reuse the same component for the same task, they may use the same set of operations. Additionally, different usages of a component often result in different frequencies of operation calls. Therefore, the frequency of operation calls for different operations of a component partially reflects its usage. In this paper, a statistical analysis is performed on the operation call frequencies of candidate examples. A vector, VOj, is used to represent the call frequencies of operations in a candidate example j, i.e., $VOj = <m_{1j}, m_{2j}, m_{3j} \ldots m_{nj}>$. Here, n represents the total number of operations in the component, and the element m_{ij} represents the number of times operation i is called in candidate example j. The similarity, $sim_{op}(x, y)$, between the operation call frequency information of two candidate examples x and y is calculated in this paper by computing the cosine value of the angle between the two vectors.

$$sim_{op}(x, y) = COS(VOx, VOy) \tag{1}$$

Then, like above, calculate the similarity of annotation information $sim_{co}(x, y)$.

Total Similarity Calculation and Code Clustering. By utilizing the similarity of operation invocation frequency, $sim_{op}(x, y)$, and the similarity of comment information, $sim_{co}(x, y)$, we can derive the overall similarity sim(x, y) between two candidate examples x and y. The formula is as follows:

$$sim(x, y) = \alpha \, sim_{op}(x, y) + (1 - \alpha) \, sim_{co}(x, y) \tag{2}$$

After completing the clustering process, it is necessary to select a representative example from each group. Here, we choose the example within the same group that has the highest similarity with the other examples.

Let G represent a specific group obtained through clustering, and let x represent a candidate example in group G.

Once the representative example is selected, the remaining candidate examples within the group will be removed from the result list as duplicate instances.

3.3 Grading Strategy

After completing the clustering and selecting representative examples from each cluster, we provide different usage examples of the components, which can help developers choose the most suitable reusable solution. However, developers often tend to only look at the top portion of the result list. Therefore, it is necessary to assign a reasonable ranking to the selected representative examples so that developers can understand the usage of the components with less effort. In our approach, the following criteria are used for ranking:

1) Number of examples in the cluster: If a cluster contains a larger number of examples, the representative example selected from that cluster is assigned a higher rank. This is because selecting a representative example from a cluster with more examples indicates a higher usage frequency, which can better represent the typical usage of the component.

2) Comment information of the examples: Examples with more comment information are assigned a higher rank. This is because more comment information in the examples helps developers understand the usage of the component.

4 Conclusion

Components are the foundation of advanced software engineering and the rising stars of the future. The emergence of component technology stems from the engineering concept of software reuse, which originated from the need to handle issues in legacy systems. Whether it is procedural, object-oriented, or component-based, the iterative and incremental nature of software processes remains unchanged.

The purpose of this study is to improve software construction technology in software reuse by focusing on component technology. Contemporary software development is no longer the same as traditional software development; it has shifted toward component-based software development. To select the most suitable component from the results of component retrieval, it is necessary to understand the different usages of each component in the retrieval results. This paper proposes the UEBCS method based on examples of component usage and addresses the challenges of component retrieval and selection from the perspective of specific clustering algorithms. This method effectively enhances the efficiency of software development and significantly reduces software development costs.

In summary, this paper starts with an overview of component technology and explores software engineering techniques based on components. It provides valuable insights into the application of component technology in software system development. It is reasonable to believe that with the continuous progress of technology, software development methods based on component-based and software reuse technologies will promote the deepening of software development toward efficiency, quality, and systemization.

Acknowledgement. This work was supported by the National Key R&D Program of China under Grant No. 2020YFB1710200.

References

1. Goguen, J.A.: Reusing and interconnecting software components. Computer **19**(02), 16–28 (1986)
2. Broy, M., Deimel, A., Henn, J., et al.: What characterizes a (software) component? Softw. Concepts Tools **19**, 49–56 (1998)
3. Bertoa, M.F., Troya, J.M., Vallecillo, A.: Measuring the usability of software components. J. Syst. Softw. **79**(3), 427–439 (2006)
4. Andrianjaka, R.M., Razafimahatratra, H., Mahatody, T., et al.: Automatic generation of software components of the Praxeme methodology from ReLEL. In: 2020 24th International Conference on System Theory, Control and Computing (ICSTCC), pp. 843–849. IEEE (2020)
5. Levy, O., Feitelson, D.G.: Understanding large-scale software-a hierarchical view. In: 2019 IEEE/ACM 27th International Conference on Program Comprehension (ICPC), pp. 283–293. IEEE (2019)
6. Comella-Dorda, S., Dean, J.C., Morris, E., Oberndorf, P.: A process for COTS software product evaluation. In: Dean, J., Gravel, A. (eds.) ICCBSS 2002. LNCS, vol. 2255, pp. 86–96. Springer, Heidelberg (2002). https://doi.org/10.1007/3-540-45588-4_9
7. Ballurio, K., Scalzo, B., Rose, L.: Risk reduction in cots software selection with BASIS. In: Dean, J., Gravel, A. (eds.) ICCBSS 2002. LNCS, vol. 2255, pp. 31–43. Springer, Heidelberg (2002). https://doi.org/10.1007/3-540-45588-4_4
8. Mancebo, E., Andrews, A.: A strategy for selecting multiple components. In: Proceedings of the ACM Symposium on Applied Computing, vol. 2005, pp. 1505–1510 (2005)
9. Lau, K.K.: Software component models. In: Proceedings of the 28th International Conference on Software Engineering, pp. 1081–1082 (2006)
10. Capilla, R., Gallina, B., Cetina, C., et al.: Opportunities for software reuse in an uncertain world: from past to emerging trends. J. Softw. Evol. Process **31**(8), e2217 (2019)
11. Li, R., Etemaadi, R., Emmerich, M.T.M., et al.: An evolutionary multiobjective optimization approach to component-based software architecture design. In: 2011 IEEE Congress of Evolutionary Computation (CEC), pp. 432–439. IEEE (2011)
12. Badampudi, D., Wohlin, C., Petersen, K.: Software component decision-making: in-house, OSS, COTS or outsourcing - a systematic literature review. J. Syst. Softw. **121**, 105–124 (2016)

13. Rotaru, O.P., Dobre, M.: Reusability metrics for software components. In: The 3rd ACS/IEEE International Conference on Computer Systems and Applications, 2005, p. 24. IEEE (2005)
14. Gill, N.S.: Importance of software component characterization for better software reusability. ACM SIGSOFT Softw. Eng. Notes **31**(1), 1–3 (2006)

A Data Security Protection Method for Deep Neural Network Model Based on Mobility and Sharing

Xinjian Zhao[1], Qianmu Li[2], Qi Wang[3,4], Shi Chen[1], Tengfei Li[2], and Nianzhe Li[2(✉)]

[1] State Grid Jiangsu Information and Telecommunication Company, Nanjing 210024, Jiangsu, China

[2] School of Computer Science and Engineering, Nanjing University of Science and Technology, Nanjing 210094, Jiangsu, China
122106010835@njust.edu.cn

[3] State Grid Smart Grid Research Institute Co., Ltd., Nanjing 210003, Jiangsu, China

[4] State Grid Key Laboratory of Information and Network Security, Nanjing 210003, Jiangsu, China

Abstract. With the rapid development of digital economy, numerous business scenarios, such as smart grid, energy network, intelligent transportation, etc., require the design and distribution of deep neural network (DNN) models, which typically use large amounts of data and computing resources for training. As a result, DNN models are now considered important assets, but at great risk of being stolen and distributed illegally. In response to the new risks and challenges brought by frequent data flow and sharing, watermarks have been introduced to protect the ownership of DNN models. Watermarks can be extracted in a relatively simple manner to declare ownership of a model. However, watermarks are vulnerable to attacks. In this work, we propose a novel label-based black-box watermark model protection algorithm. Inspired by new labels, we design a method to embed watermarks by adding new labels in the model, and to prevent watermarks from being forged, we use encryption algorithms to generate key samples. We conduct experiments on the VGG19 model using the CIFAR-10 dataset. Experimental results show that the method is robust to fine-tuning attacks, pruning attacks. Furthermore, our method does not affect the performance of deep neural network models. This method helps to solve scenarios such as "internal data circulation, external data sharing and multi-party data collaboration".

Keywords: Ownership · New Label · Encryption Algorithm · Deep Neural Network

1 Introduction

With the development of intelligence and information technology, data deep neural network models have become part of data assets and play an important role in the digital economy. Take the new power system as an example, there is extensive business interaction between the power grid and upstream and downstream units, prompting the deep learning model of power data to flow and share among third-party units, government, etc. The data model from generation to destruction is not a one-way, single-path simple flow model, and is no longer limited to internal organizational flow, but may flow from the grid to another controller. Due to the mobility and replicability of data models, once shared and distributed out, the control of data models will be quickly lost and cannot be effectively traced after model attack problems occur.

In the past few years, deep learning technology has developed rapidly and has been deployed and applied in all aspects of society. Especially with the development of computer hardware resources and the advancement of big data technology, deep learning technology has achieved important results in many challenges, including face recognition, automatic driving and natural language processing. As a key part of artificial intelligence, deep learning has brought great convenience to our lives. Training a deep neural network (DNN) with good performance often requires a lot of manpower and material resources. Including: (1) large-scale and labeled data are required, and all possible situations in the target program are covered by large-scale data. For some situations, it may be necessary to prepare ultra-large-scale data. (2) A large number of computing resources are required to adjust the model structure or model weights. The purpose is to obtain a deep neural network with excellent performance. For some scenarios, a large amount of computing resources may be required. Obviously, the deep neural network with excellent performance can only be obtained by the model owner after spending a lot of manpower and material resources. They are the crystallization of the wisdom of the model owner. Based on this, we believe that the model owner has the corresponding intellectual property rights to the deep neural network. Since training a deep neural network with excellent performance requires a lot of resources, model owners can sell the model or provide corresponding services based on the model to obtain corresponding benefits. However, many problems may be involved in this, such as the attacker obtains the model through illegal means, or even copies and sells the model, which undoubtedly damages the interests of the model owner. In view of the above considerations, we believe that it is necessary to design an intellectual property protection technology for the deep neural network model, and verify the ownership of the model in a specific way, so as to protect the interests of the model owner.

Protecting the intellectual property rights of deep neural network models has attracted worldwide attention. In May 2016, the Japanese government's "Intellectual Property Strategy Headquarters" has begun planning to protect the property rights of music and literary works created by AI. On September 16, 2022, the General Office of the State Council recently issued the "Opinions on Further Optimizing the Business Environment and Reducing the Institutional Transaction Costs of Market Entities", which clearly stated that it is necessary to continuously strengthen the protection of intellectual property rights, and improve the intellectual property protection system for new fields and new business forms such as artificial intelligence.

The concept of a black box is commonly used in fields such as technology, engineering, and data science. It represents a level of abstraction that allows users to interact with complex systems without needing to understand the intricate details of their inner workings. Instead, users can focus on the inputs and outputs, treating the black box as a "black box" with unknown internals. Black boxes are often employed when dealing with proprietary software, algorithms, or hardware components. For example, in machine learning, sophisticated models may be treated as black boxes, where users feed in input data and receive predictions without understanding the specific steps or calculations involved in generating those predictions.

A watermark is a visible or embedded piece of information that is added to an image, document, or video to indicate its authenticity, ownership, or to deter unauthorized use. Watermarks are commonly used in various industries such as photography, publishing, and digital media to protect intellectual property and prevent unauthorized copying or distribution. Watermarks can take different forms, including text, logos, patterns, or symbols, and are typically overlaid on top of the content without obstructing its visibility. They can be semi-transparent or opaque, depending on the desired level of visibility. The placement of watermarks can vary, but they are often positioned strategically to discourage tampering or removal.

In order to protect the ownership of deep neural models, we propose a black-box watermark protection method based on new labels, which embed watermarks through new labels. In our work, we make the following contributions:

1. We propose a black-box watermark protection method based on new labels, by embedding new labels to embed watermarks, which can embed watermarks without affecting the decision boundary of the original model.
2. We encrypt irrelevant samples multiple times, and set their labels as new labels to form key samples, and at the same time prevent key samples from being forged through encryption algorithms.
3. We verify the effectiveness and robustness of our method through experiments.

2 Related Research Work

The black box-based model watermarking method cannot know the specific weights of the deep neural network model, so the watermarking can only be verified according to the input and output. The black box-based model watermarking method usually completes the watermark embedding by embedding a backdoor, that is, given a target deep neural network model one or a group of special instances (often called Trigger), when the target deep neural network model performs splitting tasks, it will perform special classification tasks and classify special instances into preset labels.

Figure 1 shows an example of watermark embedding based on a black box model. The trigger is the white square in the lower right corner of the image. Key samples are constructed by adding triggers, and the label of the key sample is set to the target label "Cat". After training, the target deep neural network model can identify the key sample as the target label, and at the same time identify the correct label for other normal samples without triggers.

The article [1] uses the back door to embedding watermarks in deep neural network models. Specifically, a set of non-entangled key data is first selected, and then labels are randomly selected from all input samples and bound with key data to form key samples. This method investigates two watermarking methods, for the first method, a deep neural network model is first trained without key samples, and then the model is

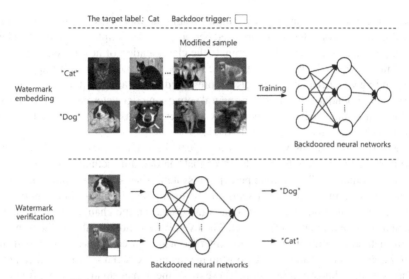

Fig. 1. Deep neural network model backdoor embedding.

retrained with key samples. The second method feeds both key samples and common samples into the model for training. To verify the watermark, the predicted labels of key samples are compared with the expected labels. For non-watermarked models, key samples are often not assigned to expected labels, and for watermarked models, key samples are often assigned to correct expected labels. The article [2] proposes a backdoor sample construction method based on text, noise and irrelevant samples, and triggers watermark verification through specific triggers. On this basis, in order to apply the backdoor technology to the embedded system, the article [3] proposes a proprietary protection method to protect the neural network model of the embedded system, using the specific information generated by it as a trigger to design the watermark.

In the above watermarking scheme based on the model backdoor, the classification of key samples with wrong labels will inevitably affect the decision boundary of the model on the original task. The article [4] proposes a new black-box-based model watermarking method. Through the boundary splicing algorithm, key samples are used to clamp the decision boundary, and the performance of the model is guaranteed under the condition of weakly affecting the decision boundary. The points chosen in the article should be close to the decision boundary of the original model, that is, their classification is not trivial and highly model-dependent, for some models it may be classified correctly, but for others it may be misclassified. In this method, adversarial perturbation is used to disturb the model by adding adversarial perturbation to the sample. Given a trained model, any well-classified example can be modified in small ways, resulting in the current misclassification. Such modified samples are called "adversarial samples", or adversaries in short. Corresponding to it is "benign adversarial example", that is, adding an adversarial perturbation to the sample will not lead to classification errors. The article [5] proposes to embed the watermark by adding a new label. This method minimizes, if not eliminates, the effect of boundary distortions by adding a new label to key samples. Instead of drawing key samples from marginal distributions in the sample space, this method considers superimposed perturbations to samples or uncorrelated natural samples as new features that dominate the classification of new classes. In theory, after adding new labels, the boundaries are not distorted and all the advantages of the corresponding watermark-free models are preserved. From another point of view, compared with the boundary-distorted deep neural network model watermarking method, this method requires the smallest number of key samples in terms of watermark embedding, ownership verification and false positive rate.

The article [6] embeds watermarks on selected samples in the frequency domain to form key samples. This method believes that the domain method is superior to the space domain method. The human perception system is sensitive to a small number of frequency bands, and some coefficients can be manipulated without being noticed; in the space representation after the inverse transformation, the change of the frequency coefficient will be diluted to each pixel. The overall distortion is subtler; the transform domain method is compatible with the international data compression standard, it is easy to implement the watermark algorithm in the compressed domain, and it can resist the corresponding lossy compression. These make the transformations in the frequency domain less visible. The article [7] proposes a way to generate trigger sets by introducing a hash function to create key samples. This method introduces a one-way hash function, trigger samples used to prove ownership must form a one-way chain, and assign corresponding labels to them. The method is applicable to various neural networks and training methods. As long as the owner follows the method to embed the watermark, they can successfully provide a valid set of triggers to prove their ownership. For an attacker who cannot train the network, although it has free access to the watermarked network, it is impossible to construct a trigger sample chain and the matching relationship between the trigger sample and the assigned label, which can make the attack invalid. The article [8] proposes a black-box based watermark embedding method that utilizes an autoencoder to embed a specific watermark into an image. Through iteration, the feature distribution of the watermarked image is made similar to that of the original image to prevent detection. The article [9] is different from the small number of key samples generated in the past, but a large number of key samples are generated. In this method, the number of key samples is the same as that of ordinary samples, and ordinary samples and key samples are used for training. Since the number of key samples is large, the watermark is difficult to remove, and adding new watermarks will degrade the performance of the model. This method first converts the picture into a scalar, and then divides the picture into blocks of the same size, and then converts each block into a one-dimensional vector and stitches them together to generate a key k. The value of the key is 0 or 1. Then the scalar is multiplied by 255 and converted to 8-bit binary, and then each scalar is operated. If the key value is 0, the value of the scalar will not change. If the key value is 1, the value multiplied by 255 and then multiplied by 2 to the L power minus 1, a large number of key samples can be generated by this method. The article [10] proposes to protect important parameters, based on the principle that the largest weight usually plays the largest role and fine-tuning and pruning always modify the parameters with lower weights. This method is first carried out on the basis of black boxes. The watermark is embedded, and then the model weights are adjusted, and the other weights are correspondingly reduced while ensuring that the maximum value in each filter remains unchanged, so as to resist the threat of fine-tuning and pruning.

Different from the above methods, the watermark proposes in [11] is deployed in the prediction API of the model, and dynamically adds watermarks to some queries by changing the prediction response of the client. The article [12] proposes a method of null embedding, which includes the watermark in the initial training of the model. Since the null embedding does not depend on incremental training, it can only be trained as a model during the initialization period, because it is difficult for the opponent to reproduce embed watermark.

3 Algorithm Design

3.1 Build Trigger

In this section we will describe how to generate key samples and the specific process of generating key samples. The generation of key samples directly affects the quality and robustness of our watermark.

We first select a one-way hash function H_x and a key k. The one-way hash function can change an input message string of any length into a fixed-length output string. At the same time, it is difficult to obtain the input string from the output string. This output string is also called hash value. In order not to affect the decision boundary of the target deep neural network model, we select a picture from other irrelevant data sets X as input, and operate on image X using our chosen one-way hash function H_x and key k. The corresponding formula is:

$$X_1 = H_x(X, k) \tag{1}$$

For formula (1), after one conversion, we can obtain the result X_1 of hashing image X using one-way hash function H_x and key k, and the hash result X_1 and the picture X have the same size and channel. For the hash operation, we do not necessarily only perform a single hash operation on the picture, we can perform X multiple hash operations on the picture, and the corresponding formula is:

$$X_l = H_x(X_{l-1}, k), l = 2, 3, \cdots, L \tag{2}$$

where L is the number of corresponding hash operations.

In our black-box watermarking method, which H_x can be any hash function, we take an image X as input and output an image of the same size X_l. In practice, we can define some standard one-way hash functions as H_x. For example, for a picture X_{l-1}, we divide it into multiple blocks, each non-overlapping block contains 32 pixels, as input we can get a block containing 32 pixels, then we use it as the block corresponding to the position of the picture X_l.

3.2 SHA256 Algorithm

SHA256 is an encrypted hash function that can perform irreversible and unique encryption transformations on data. The SHA256 algorithm is widely recognized as a secure and reliable encryption algorithm, and can also have extremely low collision probability in scenarios with large amounts of data. The hash value generated by SHA256 is 256 bits, and shorter plaintext can also generate longer hash values, making the hash value more difficult to guess and crack. The SHA256 algorithm has a relatively fast computational speed, making it suitable for large-scale data encryption processing in deep learning scenarios. It can efficiently protect the integrity and security of deep model data, and make our proposed method highly compatible and versatile. In this black-box watermarking method, we adopt the SHA256 algorithm as our one-way hash function, and in this section, we will describe the SHA256 algorithm.

SHA256 algorithm can operate on data whose length is less than or equal to 2^{64} bits, and for data less than 2^{64} bits, it can be aligned and extended to meet the conditions. Then the data is divided into n message blocks with a length of 512 bits, and n times of iterative operations are performed. This algorithm only processes one message block at a time, converts it into 64 32-bit message words, and then uses the Map function composed of logical functions and initialized constants to perform 64 iterative operations on the message words. After the operation of the last message block, we can get the final 256-bit summary.

For the SHA256 algorithm, it can be divided into six parts: message preprocessing, constant initialization, logical operation function, segmentation message, Map function iterative compression and calculation of the final message digest. We will further describe the SHA256 algorithm.

Message Preprocessing. For the input message, it is necessary to fill the bits and process the original length information. In this process, certain principles must be followed, and relevant information should be filled after the original information.

Filling Bits. After the original information is input, we need to complete the filling after the input information to ensure that the information can satisfy the remainder of 448 after dividing by 512. The rule of information filling is: for the first added bit, we need to add 1, and for other additions, we need to add 0, until the previous rule is satisfied. One thing to note is that even if the length of the input information has satisfied the remainder of 448 after dividing by 512, we still need to supplement the information. At this time, we only need to add 512 bits after the input information.

Append the Original Length. In this step, the length of the input information is expressed as 64-bit binary data, and these binary data are appended to the end of the data processed by supplementary bits, and these filled data are related to the input message itself. In memory, the encoding method of the length is big-endian data storage in reverse byte order.

After stuffing bits and appending the original length, we can get the message data after message preprocessing. At this time, the length of the message data can be divisible by 512. The purpose of this is that when the message is segmented, it can be divided into message blocks of length 512 bits.

Constant Initialization. In the operation process of SHA256, we need to define the 8 initial values of the hash digest directly required in the iterative calculation of the Map function, and also need 64 constant key used to process the message block. For the determination of the initial value of the hash digest and the constant key, there are statistical calculation rules. Specifically, we need to calculate the prime number that takes the required number from the beginning, and then calculate the square root of the hash value and the cube root of the constant key, and finally the first 32-bit value in the result we get is the final value we need.

It should be noted that the 8 initial digests are the initial hash values to be input, and the 64 initial constants are input one by one as keys in the iterative function of the Map function.

Logical Operation Function. For the SHA256 algorithm, its core is the combined operation of different logic functions, and the operations in the SHA256 algorithm are all bit operations. And 6 logic functions are used in the SHA256 algorithm, and are used in combination in the subsequent Map function iterative operation. The specific definition formulas of the six logic functions are as follows:

$$
\begin{aligned}
ch(x, y, z) &= (x \wedge y) \oplus (\neg x \wedge z) \\
ma(x, y, z) &= (x \wedge y) \oplus (x \wedge z) \oplus (y \wedge z) \\
\sum 0(x) &= S^2(x) \oplus S^{13}(x) \oplus S^{22}(x) \\
\sum 1(x) &= S^6(x) \oplus S^{11}(x) \oplus S^{25}(x) \\
\sigma_0(x) &= S^7(x) \oplus S^{18}(x) \oplus R^3(x) \\
\sigma_1(x) &= S^{17}(x) \oplus S^{19}(x) \oplus R^{10}(x)
\end{aligned}
\tag{3}
$$

The specific functions of various bit operation symbols proposed in the above formula (3) are shown in Table 1.

Segment Message. For segmenting a message, it is mainly divided into two steps. First, the message should be divided into different message blocks, and then the corresponding message words should be constructed from the message blocks. After message preprocessing, we can process the original input information into an information length divisible by 512, and then we can divide the input information into n different message blocks with a length of 512 bits. In this way, when calculating the message digest in subsequent iterations, one message block can be processed each time.

Table 1. The specific meaning of different bit operation symbols.

symbol	describe
$\neg$	bitwise negation
$\oplus$	bitwise XOR
$\wedge$	bitwise conjunction
R^n	shift right by n bits
S^n	Rotate right by n bits

In each iterative operation, there are 64 iterative compression operations of the Map function. In this process, each message block is compressed into the corresponding hash digest obtained in the iteration process. For each 512-bit message block, we can construct 64 32-bit message words through the operation specified above. In this process, the current message block is directly divided to obtain the first 16 message words, respectively W[0],W[1],$\cdots$,W[15].Through the logic function we defined above, and then perform recursive calculations through the following formula (4), the remaining 48 message words can be obtained after the calculation, and the specific calculation formula is:

$$W[t] = \sigma_1(W[t - 2]) + W[t - 7] + \sigma_0(W[t - 15]) + W[t - 16] \qquad (4)$$

Map Function Iterative Compression. In the SHA256 algorithm, the core content is the iterative compression operation of the Map function, and the iterative compression of the Map function will be executed 64 times in the SHA256 algorithm. The iterative compression operation of the Map function will process a 32-bit message word each time, and can compress the data information of a message block into the currently obtained hash summary. This is the key step for the SHA256 algorithm to realize message compression.

Specifically, for compressing a message block, 64 times of iterative compression operations of the Map function are required. In each iteration process, for the t-th Map function operation, the current $h_t(a)$, $h_t(b)$, $\cdots$, $h_t(h)$ total of 8 hash values can be operated, and then a new round of message digest is obtained $h_{t+1}(a)$, $h_{t+1}(b)$, $\cdots$, $h_{t+1}(h)$. Use the obtained new message digest as the initial value of the $(t+1)$-th Map function operation. During the whole compression process, the message words are input sequentially, and the non-linear logic functions such as ch, ma, $\sum 0$ and $\sum 1$, etc. designed by the algorithm are used for combined operations and used to process 64 different small message words. Finally, the message block is stored in the hash digest obtained in this round after undergoing 64 iterations of the Map function.

Calculate The Final Message Digest. In order to fully compress the preprocessed input information into a 256 -bit digest, the SHA256 algorithm needs to perform n rounds of iterations in order to complete the superposition of message blocks. Each round includes 64 times of Map function iterative compression operations. In the constant initialization process, we need to define 8 hash initial values, and use them as the initial message digest when compressing the first message block H_0. After the i-th round of processing, we add the content of the currently processed message block to H_i to get H_i', and add H_i and H_i', as the initial hash digest for the next iteration. After n operations, we can get n message blocks and get the final H_n. At this time, H_n is the message digest after the SHA256 algorithm.

3.3   Generate Key Samples

Taking the deep neural network model for image classification as an example, select the target deep neural network model M, which can classify the data of the $\Delta - 1$ class. Δ is represents the total number of classes. In order to generate key samples, we need to select several pictures to form a set $\mathbb{D}$. At the same time, in order not to affect the decision boundary of the target deep neural network model in our method, we do not select pictures from the original data set to generate key samples. Instead, images are selected from irrelevant sample sets to form key samples. For example, for the target deep neural network model trained using the CIFAR-10 dataset, we can extract several pictures from the CelebA face dataset to form the dataset $\mathbb{D}$.

In our method, we choose the SHA256 algorithm as the one-way hash function H_x, and we use the SHA256 algorithm to hash the image. The specific formula is as follows:

$$b\prime = SHA256(b\|k) \tag{5}$$

For formula (5), b refers to the block in X_{l-1}, b' refers to the block in X_l, X_{l-1} refers to the picture obtained after $l - 1$ times of hash operations, X_l refers to the picture obtained after l times of hash operations. For each hash operation, we append the key k to b. For the SHA256 algorithm, we can accurately divide the output into 32 pixel values. This is why we divide 32-pixel values into a block. Specifically, the output of the SHA256 algorithm is a hexadecimal number with a length of 64 bits. If it is converted into binary, we can obtain a length of 256 bits of data. For pictures, the values in the RGB channel are all 0 ~ 255, at this time we use 8-bit data to represent the values, so for the data generated by the SHA256 algorithm, we can represent it as a block. Figure 2 shows the initial image, and the corresponding five trigger images.

All the pictures in the set $\mathbb{D}$ are processed by the SHA256 algorithm, we can obtain a new collection of pictures $\mathbb{D}_1$. For the new set $\mathbb{D}_1$, we set its label as the label of class Δ, and finally we can obtain the key sample set $\mathbb{P} = \{p_i, \Delta\}_{i=1}^{n}$, among $n = |\mathbb{D}_1|$. After we get the key sample set, we mix the key sample set with the original data set and use it as the training set to train the target deep neural network model. When the performance of the target deep neural network model reaches the standard, we can think that a watermark is embedded in the target deep neural network model, that is, we have obtained a model with an embedded watermark M'.

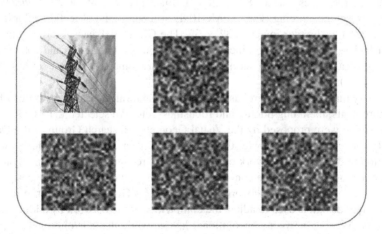

Fig. 2. Initial image and corresponding trigger image.

3.4 Watermark Verification

The black-box-based deep neural model watermark is different from the white-box-based deep neural network model watermark. The black-box watermark cannot know the specific structure and weight of the deep neural network model, that is to say, we can only use the input and output of the deep neural network model to verify the watermark.

When we find that a deep neural network model is highly similar to the deep neural network model we embed the watermark in, we can randomly select several pictures from the key sample set $\mathbb{P}$ to form a subset $\mathbb{P}_1$ for verifying the watermark, and then we input the selected verification set $\mathbb{P}_1$ into a highly similar deep neural network model, and observe the output of the deep neural network model to judge whether the output label is the label of the class Δ.

The watermark can be judged in this way because we add a new label to the target deep neural network model, and the label $\mathbb{P}$ of the key sample set can be predicted as the class Δ, but for other models without embedded watermarks F, when we set the key sample set $\mathbb{P}$ input into the model F for prediction, the prediction result has no relationship with the class label Δ. The specific formula is as follows:

$$\text{Prob}[F(x) = \Delta] \equiv 0 \tag{6}$$

Among them $x \in \mathbb{P}_1$.

4 Experiment

To verify the effectiveness of our proposed robust black-box-based watermarking, we employ the popular CIFAR-10 dataset. The CIFAR-10 [13] dataset is often used to train machine learning and computer vision. It is a subset of the color dataset with labels and the number of samples reaches 80 million, and it is also one of the most widely used datasets in machine learning. The CIFAR-10 dataset has a total of 60,000 color images, and the size of each color image is 32×32. The CIFAR-10 dataset has 10 classes in total, each class has 6000 images, including 5000 pictures for training and 1000 pictures for testing, that is, the training set has 50000 pictures images, the test set has 10,000 images (Fig. 3).

To verify that our proposed robust black-box-based watermarking can satisfy multiple requirements, including fidelity and robustness, etc. We selected the VGG series of models, which were proposed by the Visual Geometry Research Group of the Department of Science and Engineering, Oxford University in 2014. Due to the successful application of deep neural network models in image recognition tasks, Simonyan et al. proposed a simple but effective convolutional neural network architecture design principle, which they named the network architecture as VGG. In this design principle, a modular layer design is used to adjust the complexity of the network by adding smaller convolutional layers between convolutional layers, while the convolutional layers can also learn linear combinations of structural feature maps. In order to optimize the network structure, the designer added a fully connected layer after the convolutional layer to maintain the spatial resolution. In the 2014-ILSVRC competition, the VGG network

plane	car	bird	cat	deer	dog	frog	horse	ship	truck

Fig. 3. CIFAR-10 dataset example.

showed good results on both image classification and localization problems, while it also maintained a simple and homogeneous topology.

The black-box watermark protection method based on the new label proposed in this article is implemented on the Ubuntu 18.04 server. This method uses an RTX3090 graphics card with a memory size of 20G. All experimental codes are written based on PyTorch and Python, the version of PyTorch is $1.7.0 + cu110$, and the version of Python is 3.6.9.

For the generation of the key set, in this experiment, we randomly select 30 face pictures from the celebA face dataset, use the SHA256 algorithm to encrypt each face picture for 5 rounds to form an encrypted image, and finally the label of the encrypted image is set to a label that does not exist in the original model to form a key sample. In the training of the watermarking model, we mix key samples with the training set to form a new training set without making any changes to the training set of the original model.

4.1 Fidelity

In this section, we will verify the fidelity. In order to ensure fairness, the watermark model and the original model are trained. Except for the difference between the training set and the classification category, other conditions are similar. In this experiment, we will conduct 100 rounds of training on both the original model and the watermark model, and observe the accuracy of the original model and the watermark model under different training rounds in the test set, to observe whether the black box watermark model protection method based on the new label will have an impact on the performance of the model.

As shown in Fig. 4, it can be found that the accuracy of the original model and watermark model increases with the number of training rounds. In the previous rounds of training, the accuracy of the original model and watermark model increases with the number of rounds. After a rapid increase, the accuracy of the original model and the watermark model basically reached the highest level when the number of training rounds reached 60 rounds, and maintained at a high level in the end. After 100 rounds of training, the performance of the original model is 92.65%, while the performance of the

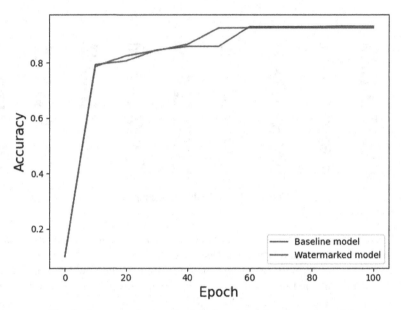

Fig. 4. Accuracy change of original model and watermark model.

watermark model is 93.17%. It can be found that the increase of key samples not only does not cause the performance of the watermark model to decline, but even has a slight increase. An increase of 0.52% compared to the original model. Through experiments, we can find that the black-box watermarking model protection method based on the new label does not have a negative impact on the performance of the model.

4.2 False Detection

In the experiments in this subsection, we will measure the false detection of the watermarking model. In general, black-box watermarking, there is no need to verify the misdetection of the watermarking model, because even some normal samples are identified as key samples will not have a serious impact on the verification of the watermark. In the new label-based black-box watermarking model protection method, we want to minimize or avoid false positives, because for other black-box methods, other samples are misidentified as key samples and will not reveal other information, but in In the black-box watermarking model protection method based on new labels, if other irrelevant samples are misidentified as key samples, the information of our newly added labels may be leaked, which may have a certain negative impact on our watermarking model.

In this experiment, neither the common samples in the training set nor the common samples in the test set will be misidentified as key samples, which means that in general, we don't need to worry about other irrelevant samples being identified as key samples. That is to say, there is no need to worry about the leakage of our watermark information.

After training, the accuracy rate of key samples on the watermark model is 100%, that is to say, the accuracy rate of the black-box watermark model protection method based on new labels can achieve the best effect on key samples. At the same time, other irrelevant samples will not be misidentified as key samples, which means that our new label-based black-box watermarking model protection method can successfully protect the ownership of the model and will not wrongly verify the watermark, so we can say that our method is efficient and reliable.

4.3   Defense Against Fine-Tuning Attacks

In this experiment, we will discuss the robustness of the watermark model against fine-tuning attacks, which means that when the model is modified, whether the model modification will have a significant negative impact on the watermark model, that is, whether the verification accuracy of the watermark model for key samples can be maintained at a high level to verify the ownership of the watermark model. The black-box watermarking protection method based on new labels is different from the robust watermarking algorithm based on the fusion of multi-pruning methods. The watermark embedded in the black-box watermarking protection method based on new labels is embedded in new labels, so after the watermark model is generated, the watermark embedded by this method cannot be deleted without changing the deep neural network model.

In this fine-tuning attack experiment, we divide the test set into two parts, the first part of the test set is the first 80% of the test set as the new training set, and the latter part of the test set is the last 20% of the test set as a new test set, we fine-tune the watermark model based on this, and count the accuracy of the watermark model for key sample verification after fine-tuning.

Table 2. Key sample detection accuracy of watermarking model.

Number of rounds	Accuracy
1 0	100%
3 0	100%
5 0	100%
7 0	100%
9 0	100%
100	100%

As shown in Table 2, it can be found that even after 100 rounds of fine-tuning attacks on the watermark model, the accuracy of the watermark model generated by using the black-box watermark protection method based on the new label is still 100% on key samples, and the loss value is 0%. This is because the black-box watermarking protection method based on the new label proposed by us is to embed the watermark through the new label, that is to say, it will not affect the decision boundary of the original model, so even if the fine-tuning attack on the watermark model has a partial impact on the decision boundary of the watermark model, it will not have a large impact on the decision boundary of the new label. Therefore, the detection accuracy of the watermark model for key samples has always been 100%. That is to say, the watermark model generated by our new label-based black-box watermark protection method has excellent robustness against fine-tuning attacks.

4.4 Defense Against Pruning Attacks

The purpose of resisting pruning attacks is to measure whether the black box watermark model protection method based on the new label is robust to the pruning method, that is, whether the watermark accuracy on the watermark model can reach the level used for watermark verification after the watermark model is pruned. Since the deep neural network model often has many parameters and complex network structure, the model often requires a large space storage and a large overhead to calculate. In order to reduce the storage space and calculation overhead, the model can be pruned to eliminate redundant parameters in the model, so that the performance of the model does not deteriorate under the condition of reducing the model.

In this experiment, we choose the pruning method proposed in article [14] to attack the watermark model, use the pruning attack to prune the watermark model, use the pruned watermark model to detect key samples, and check the accuracy.

As shown in Table 3, it can be found that for the watermark model generated by the black box watermark model protection method based on the new label, even when the pruning rate is 80%, the detection accuracy of the watermark model for key samples has always been as high as 100%. Although the detection accuracy of the watermark model for key samples has significantly decreased when the pruning rate is as high as 90%, and the accuracy rate is only 23.33% at this time. It is no longer possible to verify the ownership of the watermark model through the form of key samples, but when the watermark model is pruned by 90%, the performance of the watermark model has decreased to a level where it cannot be used. So it can be said that our proposed black box watermark model protection method based on new tags has excellent robustness to pruning attack.

Table 3. The detection accuracy of this method under the pruning attack.

Pruning rate	Accuracy
10%	100%
20%	100%
30%	100%
40%	100%
50%	100%
60%	100%
70%	100%
80%	100%
90%	23.33%

5   Experiment

This article proposes a new label-based black-box watermark model protection algorithm, which can embed watermarks in deep neural network models, while maximizing the concealment and robustness of watermarks, and reducing the impact of watermark embedding on deep neural network models. In order not to have too much negative impact on the watermark model, we add a new label on the basis of the training set of the original model. The addition of the new label will not affect the decision boundary of the original model, which can minimize the impact on the deep neural network. In order to reduce the probability of watermark forgery, we choose to encrypt key samples multiple times. Experiments show that the method proposed in this article does not affect the performance of the original model, and has good robustness against pruning attacks and fine-tuning attacks.

References

1. Adi, Y., Baum, C., Cisse, M., et al.: Turning Your Weakness Into a Strength: Watermarking Deep Neural Networks by Backdooring (2018)
2. Zhang, J., Gu, Z., Jang, J., et al.: Protecting Intellectual Property of Deep Neural Networks with Watermarking, pp.159–172 (2018)
3. Guo, J., Potkonjak, M.: Watermarking deep neural networks for embedded systems. In: 2018 IEEE/ACM International Conference on Computer-Aided Design (ICCAD), pp. 1–8. IEEE (2018)
4. Merrer, E.L., Perez, P., Trédan, G.: Adversarial frontier stitching for remote neural network watermarking. Neural Comput. Appl. **32**, 9233–9244 (2017)
5. Xiang, Y., Gao, L., Zhang, J., et al.: Protecting IP of deep neural networks with watermarking: a new label helps. In: Pacific-Asia Conference on Knowledge Discovery and Data Mining (2020)
6. Li, M., Zhong, Q., Zhang, L.Y., et al.: Protecting the intellectual property of deep neural networks with watermarking: the frequency domain approach. In: 2020 IEEE 19th International Conference on Trust, Security and Privacy in Computing and Communications (TrustCom). IEEE (2020)
7. Zhu, R., Zhang, X., Shi, M., et al.: Secure neural network watermarking protocol against forging attack. EURASIP J. Image Video Process. **2020**(1), 37 (2020)
8. Li, Z., Hu, C., Zhang, Y., et al.: How to prove your model belongs to you: a blind-watermark based framework to protect intellectual property of DNN (2019)
9. Aprilpyone, M.M., Kiya, H.: Piracy-Resistant DNN Watermarking by Block-Wise Image Transformation with Secret Key (2021)
10. Namba, R., Sakuma, J.: Robust Watermarking of Neural Network with Exponential Weighting. Computer and Communications Security (2019)
11. Szyller, S., Atli, B.G., Marchal, S., et al.: Dawn: dynamic adversarial watermarking of neural networks. In: Proceedings of the 29th ACM International Conference on Multimedia, pp. 4417–4425 (2021)
12. Li, H., Wenger, E., Shan, S., et al.: Piracy resistant watermarks for deep neural networks (2019). arXiv preprint arXiv:1910.01226
13. Krizhevsky, A., Hinton, G.: Learning Multiple Layers of Features from Tiny Images. Handbook of Systemic Autoimmune Diseases, vol. 1, no. 4 (2009)
14. Han, S., Pool, J., Tran, J., et al.: Learning both weights and connections for efficient neural networks (2015). arXiv preprint arXiv:1506.02626

A Method for Small Object Contamination Detection of Lentinula Edodes Logs Integrating SPD-Conv and Structural Reparameterization

Qiulan Wu[1] , Xuefei Chen[1] , Suya Shang[1], Feng Zhang[1(✉)], and Wenhui Tan[2(✉)]

[1] School of Information Science and Engineering, Shandong Agricultural University,
Tai'an 271018, China
zhangfeng@sdau.edu.cn
[2] Tai'an Service Center for Urban Comprehensive Management, Tai'an 271018, China
Sdtatwh@163.com

Abstract. A small object contamination detection method (SRW-YOLO) integrating SPD-Conv and structural reparameterization was proposed to address the problem of the difficulty in the detection of small object contaminated areas of Lentinula Edodes logs. First, the SPD (space-to-depth)-Conv was used to improve the MP module to enhance the learning of effective features of Lentinula Edodes log images and prevent the loss of small object contamination information. Meanwhile, RepVGG was introduced into the ELAN structure to improve the efficiency and accuracy of inference on the contaminated regions of Lentinula Edodes logs through structural reparameterization. Finally, the boundary regression loss function was replaced with the WIoU (Wise-IoU) loss function, which focuses more on ordinary-quality anchor boxes and makes the model output results more accurate. In this study, the measures of Precision, Recall, and reached 97.63%, 96.43%, and 98.62%, respectively, which are 4.62%, 3.63%, and 2.31% higher compared to those for YOLOv7. Meanwhile, the SRW-YOLO model detects better compared with the current advanced one-stage object detection model.

Keywords: Lentinula Edodes logs · contamination detection · structural reparameterization · small object detection

1 Introduction

Lentinula Edodes logs are important carriers for Lentinula Edodes cultivation and are often contaminated by sundry bacteria during the cultivation process [1–3], thus causing significant economic losses to the enterprise. At present, the contamination status of Lentinula Edodes logs still relies on manual inspection. Manual inspection is not only high in labor cost and low in efficiency, but it also requires high professional quality of inspectors and usually can only detect contamination of Lentinula Edodes logs when it is more obvious. How to detect the initial contamination of Lentinula Edodes logs in a timely and accurate manner is crucial to prevent further spread and generation of contamination and improve the quality and yield of Lentinula Edodes.

© The Author(s), under exclusive license to Springer Nature Singapore Pte Ltd. 2024
H. Jin et al. (Eds.): GPC 2023, LNCS 14503, pp. 30–46, 2024.
https://doi.org/10.1007/978-981-99-9893-7_3

In recent years, with the development of deep learning theory [4, 5], deep learning-based detection algorithms have been widely used because of their good generalization ability and cross-scene capability [6, 7]. Therefore, the use of deep learning techniques to process crop disease images has gradually become a popular research topic [8–10]. Zu et al. [11] used a deep learning method to identify contamination of Lentinula Edodes logs for the first time and proposed an improved ResNeXt-50 (32 × 4d) model for Lentinula Edodes log contamination identification. The method improves the model by fine-tuning the six fully connected layers in the ResNeXt-50 (32 × 4d) model to improve the accuracy of Lentinula Edodes log contamination recognition, thereby breaking the situation of relying on manual detection with low efficiency and easy selection errors. However, the complex network structure and low detection efficiency of this method are not conducive to deployment to mobile or embedded devices. Zu et al. [12] proposed Ghost-YoLoV4 for Lentinula Edodes log contamination identification, which used a lightweight network, GhostNet, instead of a backbone feature extraction network to ensure identification accuracy while improving real-time performance and identification speed. Although scholars have achieved certain results in Lentinula Edodes log contamination detection using deep learning techniques, the detection effect of existing studies is not satisfactory in the early stage of Lentinula Edodes logs contamination due to small contamination areas, which require small object detection. The difficulty of small object detection has been an important problem faced by object detection algorithms, and many scholars have conducted in-depth studies for this purpose [13–17]. However, no relevant literature on deep learning for small object Lentinula Edodes log contamination detection has been found in previous studies. Wang et al. [18] proposed the YOLOv7 model with faster speed and higher accuracy using the COCO dataset. YOLOv7 contains multiple trainable freebie packages that allow the detector to greatly improve detection accuracy without increasing the inference cost. In this study, we improved the YOLOv7 algorithm and proposed a model (SRW-YOLO) applicable to small object Lentinula Edodes log contamination detection. First, SPD-Conv was introduced in the MP module of the network to highlight small object contamination object features. Then, the ELAN structure in the backbone network was reparameterized using RepVGG to further improve the detection of small object Lentinula Edodes log contamination. Finally, the target location was regressed using the WIoU loss function to pay more attention to the ordinary-quality anchor boxes in order to improve the overall performance of the detection of bacteriophage contamination condition.

2 Materials and Methods

2.1 Data Acquisition

The dataset used in this study was obtained from the self-built database of the Smart Village Laboratory of Shandong Agricultural University and collected from a factory culture shed in Shandong Province. In the Lentinula Edodes log culture shed, LED strip lights were installed at intervals to ensure good lighting conditions while not affecting the culture environment of Lentinula Edodes logs. The acquisition equipment was divided into a Canon EOS 600D camera and an IQOO8 cell phone, with image resolution ranging from 1900 to 4000 in width and from 2200 to 4000 in height. Based on the collected images of Lentinula Edodes logs, the Lentinula Edodes logs were divided into Normal Lentinula Edodes logs, Aspergillus flavus Lentinula Edodes logs, and Trichoderma viride Lentinula Edodes logs, including small object contaminated Lentinula Edodes logs, as shown in Fig. 1. A total of 3156 images were collected, including 1734 images of normal Lentinula Edodes logs 700 images of Aspergillus flavus-contaminated Lentinula Edodes logs and 722 images of Trichoderma viride-contaminated Lentinula Edodes logs.

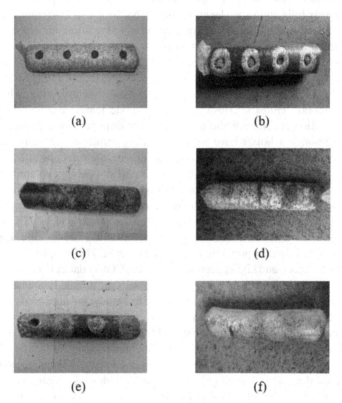

(a)   (b)

(c)   (d)

(e)   (f)

Fig. 1. Example images of Lentinula edodes logs. (a, b) normal, (c, d) Aspergillus flavus-contaminated, and (e, f) Trichoderma viride-contaminated.

2.2 Data Pre-processing

In deep learning, models need to be trained using large amounts of data in order to avoid overfitting. The adequacy and comprehensiveness of the data used are crucial to improve the accuracy of the proposed model. In order to expand the sample size, this study used data augmentation. The data enhancement strategy included various morphological operations, such as angle rotation, saturation adjustment, exposure adjustment, image flipping up and down, and random cropping, as shown in Fig. 2. By applying these operations, more samples could be generated, thus improving the generalization ability and robustness of the model. The augmented data samples included 2988 images of normal Lentinula Edodes logs, 1912 images of Aspergillus flavus-contaminated Lentinula Edodes logs, and 1512 images of Trichoderma viride-contaminated Lentinula Edodes logs, totaling 6412 images, which were saved separately in the jpg format.

Meanwhile, labeling was used as an image annotation tool to classify the types of annotations into Normal, Aspergillus flavus, and Trichoderma viride, and the label files were saved in the yolo format. Finally, the dataset was divided into a training set, a validation set, and a test set according to the ratio of 8:1:1. Among these datasets, there were 5130 images in the training set, 641 images in the validation set, and 641 images in the test set.

Fig. 2. Renderings of data enhancements.

2.3 SRW-YOLO Model Construction

In this study, an SRW-YOLO network model was designed for the problem of small object Lentinula Edodes log contamination detection, as shown in Fig. 3. Firstly, the MP module was improved by SPD-Conv to enhance the learning of small object features in the Lentinula Edodes log images and avoid the loss of fine-grained information. Secondly, RepVGG was introduced into the ELAN structure, and the structure was reparameterized to decouple the multi-branch structure and inference ordinary structure during model training, which further improved the efficiency and accuracy of inference for small object contaminated regions. Finally, the original boundary regression loss function was replaced with the WIoU loss function, which weakens the influence of high-quality anchor boxes and low-quality sample features and focuses on ordinary-quality anchor boxes, making the model output results more accurate.

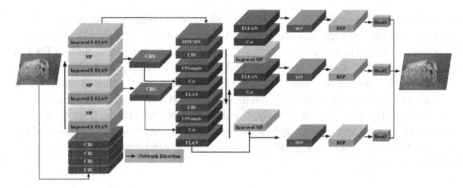

Fig. 3. The network structure of SRW-YOLO.

MP Module Based on SPD-Conv. YOLOv7 uses an MP structure to downsample the input. Downsampling is usually implemented using convolutional layers, pooling layers, or convolution with a step size greater than 1 to gradually reduce the spatial size of the input tensor and, thus, increase the perceptual field of the network. However, in the process of downsampling, it was easy to cause the resolution of Lentinula Edodes log images to decrease too fast, which would lead to a loss of information about the location and size of Lentinula Edodes log contamination, thus reducing the accuracy of detection. Therefore, to solve this problem, the MP module was improved by introducing SPD-Conv [19]. SPD-Conv consists of a space-to-depth (SPD) layer and a non-stride convolutional layer. The SPD layer slices an intermediate feature map X(S,S,C_1) into a series of sub-maps f(x,y) by downsampling the feature maps inside the convolutional neural network and the entire network.

$$f_{0,0} = X[0 : S : scale, 0 : S : scale], f_{1,0} = X[1 : S : scale, 0 : S : scale], \ldots,$$
$$f_{scale,0} = X[scale - 1 : S : scale, 0 : S : scale] \tag{1}$$

$$f_{0,1} = X[0 : S : scale, 1 : S : scale], f1, 1, \ldots,$$
$$f_{scale-1,1} = X[scale - 1 : S : scale, 1 : S : scale]; \tag{2}$$

$$\vdots$$

$$f_{0,scale-1} = X[0 : S : scale, scale - 1 : S : scale], f_{1,sclae-1}, \ldots,$$
$$f_{sclae-1,sclae-1} = X[scale - 1 : S : scale, scale - 1 : S : scale] \tag{3}$$

Given any (original) feature map $X, f_{x,y}$ which consists of the feature map $X(i, j)$ is composed of the region where $i + x$ and $j + y$ are divisible by the scale.

Thus, each subsample is mapped down by a scale factor X. Finally, the sub-feature maps are stitched along the channel dimension to obtain a feature map X. Adding a non-stride convolution after the SPD feature transformation preserves all the discriminative feature information as much as possible, and the SPD-Conv structure is shown in Fig. 4.

A total of five MP modules were constructed in the original model for the backbone network and the feature fusion network. Since there is a convolution of step 2 in the

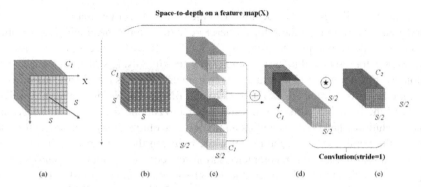

Fig. 4. Illustration of SPD-Conv when scale = 2.

second branch of the MP module, this study used SPD-Conv to replace the convolution of step 2 in the MP of the feature fusion network, as shown in Fig. 5. Considering the large input image pixels, the number of parameters, and the computational efficiency of the model, all convolutions with step size 2 in the network were not replaced in this study.

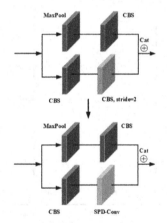

Fig. 5. Improvement of MP module.

RepVGG-based Efficient Aggregation Network Module. The efficient aggregation network module proposed in the original model is mainly divided into an ELAN [20] structure and an E-ELAN structure. The ELAN uses a special jump connection structure to control the longest gradient path, and the deeper network can learn and converge efficiently. The E-ELAN is an expansion, channel rearrangement, and transition layer architecture without destroying the original gradient path of the ELAN or changing the merging bases to enhance the learning ability of the network. However, the efficient aggregation network module may assign some important information to different groups and affect model performance. In addition, this module uses fewer convolutional layers, which can be challenging when dealing with the task of detecting small object contaminated areas of Lentinula Edodes logs. Therefore, in this study, the efficient aggregation network module was improved using RepVGG [21]. RepVGG decouples the training multi-branch topology and inference single-way structure using structural reparameterization, as shown in Fig. 6. The structural reparameterization is mainly divided into two steps: the first step is mainly to fuse Conv2d and BN (Batch Normalization) as well as to convert the branches with only BN into one Conv2d, and the second step fuses the 3x3 convolutional layers on each branch into one convolutional layer; this structure can increase the nonlinearities of the model while reducing the computation during inference. At the same time, the reparameterization reduces the computation and memory usage, which helps to handle small object contamination detection tasks. The specific improvement in this study is to introduce the RepVGG module in all ELAN structures in the backbone network.

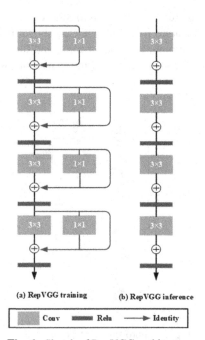

(a) RepVGG training   (b) RepVGG inference

| Conv   Relu ⟶ Identity |

Fig. 6. Sketch of RepVGG architecture.

Boundary Regression Loss Function. In an object detection task, the bounding box regression loss function is critical to the performance of the model used. The role of the bounding box regression loss function is to measure the difference between the model-predicted bounding box and the true bounding box, which affects the detection effectiveness of the model. Low-quality samples, such as small object contamination, exist in the dataset of Lentinula Edodes logs, and the geometric factors, such as distance and aspect ratio, taken into account by the traditional bounding box loss function will aggravate the penalty of low-quality examples, which may reduce the generalization performance of the model. Therefore, in this study, WIoUv3 [22] was used as the boundary regression loss function for the model. WIoUv3 proposes outliers instead of IoU to evaluate the quality of anchor boxes and provide a sensible gradient gain allocation strategy. This strategy reduces the competitiveness of high-quality anchor boxes while minimizing harmful gradients generated by low-quality examples, which contributes to the speed of model convergence and the accuracy of inference, thus improving the overall performance of model detection. This is achieved by assigning outlier β an appropriate gradient gain depending on its size, with smaller or larger outliers β being assigned smaller gradient gains that are more focused on ordinary-quality anchor boxes, with outlier β being defined as follows:

$$\beta = \frac{L_{IoU}^*}{\overline{L_{IoU}}} \in [0, +\infty) \tag{4}$$

where L_{IoU}^* is the monotonic focus factor and $\overline{L_{IoU}}$ is the sliding average of the momentum of m.

Distance attention was also constructed based on the distance metric, and a WIoUv1 with two layers of attention mechanisms was constructed as follows:

$$L_{WIoUv1} = R_{WIoU} L_{IoU}$$

$$R_{WIoU} = exp\left(\frac{(x-x_{gt})^2 + (y-y_{gt})^2}{\left(W_g^2 + H_g^2\right)^*}\right) \tag{5}$$

where L_{IoU} is the degree of overlap between the prediction box and the real box; (x, y) is the center coordinate of the predicted box; (x_{gt}, y_{gt}) is the center coordinate of the real box; and W_g and H_g are the length and width of the real box and the predicted box, respectively.

At this point, applying the outlier degree to L_{WIoUv1} obtains L_{WIoUv3}:

$$L_{WIoUv3} = rL_{WIoUv1}, r = \frac{\beta}{\delta\alpha^{\beta-\delta}} \tag{6}$$

where L_{WIoUv1} is the attention-based boundary loss, and δ with α is the hyperparameter.

When the outlier degree of the anchor box satisfies $\beta = C$ (C is a constant value), the anchor box will obtain the highest gradient gain. Since $\overline{L_{IoU}}$ is dynamic, the quality classification criteria of the anchor boxes are also dynamic, which allows WIoUv3 to construct a gradient gain allocation strategy that best fits the current situation at each moment.

2.4 Model Training and Evaluation

Model Training. In this study, SRW-YOLO used the default hyperparameters of YOLOv7. The learning rate was set to 0.01, SGD was selected for parameter optimization, and the learning rate momentum was set to 0.937. Meanwhile, a pre-trained model was used for training assistance, which could help the model achieve better initial performance. The configuration of the experimental environment in this study is shown in Table 1.

Table 1. Experimental environment configuration.

Experimental Environment	Details
Programming language	Python 3.8.10
Operating system	Linux
Deep learning framework	Pytorch 1.11.0
CUDA Version	11.3
GPU	RTX 3090
CPU	Intel(R) Xeon(R) Platinum 8358P @ 2.60GHz

Model Evaluation. To verify the performance of Lentinula Edodes log contamination detection, Precision, Recall, mAP, and FPS were used for evaluation in this study. The calculation equations are as follows.

$$Precision = \frac{TP}{TP+FP} \tag{7}$$

$$Recall = \frac{TP}{TP+FN} \tag{8}$$

$$AP = \int_0^1 P_{(r)}dr \tag{9}$$

$$mAP = \frac{\sum_{i=1}^{C} AP_i}{C} \tag{10}$$

where *TP* indicates that the target is a certain type of Lentinula Edodes logs and the network model detection also indicates a certain type of Lentinula Edodes logs. *FP* indicates that the target is not a type of Lentinula Edodes logs, but the network model detects a type of Lentinula Edodes logs. *FN* indicates that the target is a certain type of Lentinula Edodes logs, but the network model detection indicates it is not a certain type of Lentinula Edodes logs. *AP* is the area enclosed by Precision and Recall on the curve. *mAP* is the average of all categorized *AP* values; when IoU is set to 0.5, it is mAP@0.5, and mAP@0.5:0.9 means that the IoU threshold is between 0.5 and 0.9.

3 Results and Discussion

3.1 Model Visualization Analysis

After the training of the model, the feature extraction results of the first convolutional layer, the backbone module, and the last convolutional layer were visualized and analyzed in this study using class activation mapping (CAM) [23]; the information of interest to the network model can be seen from the visualized feature map. This study randomly selects an image of small target Lentinula Edodes logs contamination from the training set to visualize its characteristics. The red box area is the area contaminated by the Lentinula

Edodes log. The visual analysis results are shown in Fig. 7. The figure shows the feature visualization images of the three improvement strategies of SPD-Conv, RepVGG, and WIoUv3 regression loss function and the three stages of the SRW-YOLO comprehensive improvement model. The three stages are the first convolution layer and the feature extraction backbone. Layer and the last convolutional layer. The darker the red part, the more the model pays attention to this part of the image; this is followed by the yellow part. The bluer the heat map is, the more the model considers this part as redundant information.

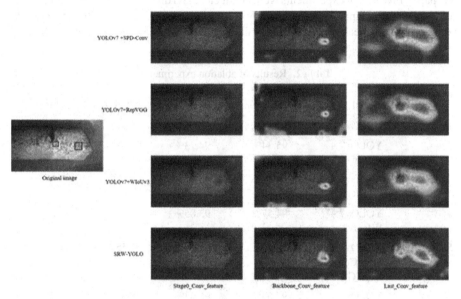

Fig. 7. Visualization of the feature map.

As can be seen from the first layer convolutional feature map, the three improvement strategies mainly focus on the low-level features of the Lentinula Edodes logs, such as edges and textures. The feature map of the feature extraction backbone convolutional layer shows more advanced feature attention, and the focus is more localized. SRW-YOLO accurately locates small target contaminated areas, and the three improvement strategies all focus on the contaminated areas of the bacterial sticks relatively accurately. However, the three improvement strategies all focus on more background redundant information to varying degrees. It can be seen from the feature map of the last convolutional layer that the features extracted by different improvement strategies are more abstract and refined, revealing how the model focuses on discriminative features in the final stage. The above improvement strategies ultimately focused on two contaminated areas. However, SPD-Conv paid too much attention to the two areas and considered more redundant pixels; Rep-Conv and WIoU3 also paid too much attention to the right areas. The feature extraction ability of the contaminated area below is weak; while SRW-YOLO focuses on key pixel areas and is more accurate. It can be observed from the feature maps from the backbone module to the last layer that the algorithm model proposed in this study plays a good role in reinforcing the feature maps, suppressing unnecessary features, and enabling better extraction of small object contamination feature information from the images.

3.2 Analysis of Experimental Results

To verify the positive impact of the improvement strategy proposed in this study on the network, ablation experiments were conducted on the Lentinula Edodes log dataset in this paper. Five sets of experiments were conducted, and different improvement modules were added for comparison with YOLOv7, with Precision, Recall, mAP@0.5, and FPS being used as the measures. The results of the ablation experiments are shown in Table 2.

Table 2. Results of ablation experiments.

Number	Model	Precision	Recall	mAP@0.5	FPS
1	YOLOv7	93.01%	92.80%	96.31%	56
2	YOLOv7 + SPD-Conv	95.34%	94.73%	96.88%	52
3	YOLOv7 + RepVGG	94.78%	93.31%	96.47%	53
4	YOLOv7 + WIoUv3	94.57%	93.85%	96.49%	64
5	YOLOv7 + SPD-Conv + RepVGG	96.16%	96.05%	97.55%	51
6	YOLOv7 + SPD-Conv + RepVGG + WIoUv3	97.63%	96.43%	98.62%	62

As can be seen from the above table, Experiment 1 provides the detection results of the original YOLOv7 network. In Experiment 2, Precision, Recall, and mAP@0.5 improve by 2.33%, 1.93% and 1.97%, respectively. This indicates that during the model downsampling process, SPD-Conv effectively alleviates the impact of the rapid decrease in resolution of the mushroom stick image, thereby strengthening the learning of effective feature representation of the Lentinula Edodes logs image and avoiding the loss of fine-grained information.In Experiment 3, Precision, Recall and mAP@0.5 improve by 1.77%, 0.51% and 1.56%, respectively. This indicates that RepVGG uses structural re-parameterization to decouple training of multi-branch topology and inference single-channel structures, which can reduce the computational complexity and memory usage of model inference while improving the efficiency and accuracy of inference on Lentinula Edodes logs contaminated areas. In Experiment 4, YOLOv7 improves Precision by 1.56%, Recall by 1.05% and mAP@0.5 by 1.58% over the YOLOv7 algorithm after using WIoUv3 as the boundary regression loss function of the network. This indicates that when YOLOv7 adopts WIoUv3, through a wise gradient gain allocation strategy, the model is more focused on ordinary-quality Lentinula Edodes logs detection anchor boxes, making the model output results more accurate. In Experiment 5, the Precision improves by 3.15% and mAP@0.5 improves by 2.64% over the YOLOv7 algorithm. This shows that when the SPD-Conv module and RepVGG module are introduced into the original network, the network inference efficiency is improved while avoiding the loss of location and size information of bacteriophage contamination, which in turn improves the accuracy of detection. Experiment 6 integrated the above improved methods, and it can be clearly seen that the detection effect is the best. Precision reaches 97.63%, which is 4.62% better than YOLOv7; Recall reaches 96.43%, 3.63% higher than YOLOv7; and mAP@0.5 reaches 98.62%, which is 2.31% better than YOLOv7. At the same time, it also maintains good real-time detection, which can meet the requirements of small object Lentinula Edodes log contamination image detection.

The ablation experiments can only verify the effectiveness of the improved strategy in this study relative to the original algorithm, but whether it can reach the leading level in different models still needs further proof. Therefore, under the same experimental conditions, a series of comparative experiments were conducted in this study to compare the performance of the improved method with the current mainstream one-stage object detection method using the Lentinula Edodes log dataset.

A comparison of the training results of different models is shown in Fig. 8. From the figure, it can be seen that the value of mAP@0.5 of the improved algorithm in this study is significantly higher than the other three models.

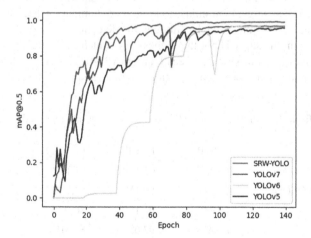

Fig. 8. Comparison of training box_loss curves of different models.

Figure 9 presents a comparison of the regression loss curves for different models with training time. After 40 iterations, the loss curves of different models gradually and steadily converge. It can be seen that YOLOv6m has poor loss convergence in this dataset and YOLOv5l has an overfitting problem after 100 training iterations. YOLOv5l and YOLOv6m are much less effective than YOLOv7 in terms of regression loss. The model proposed in this study shows a better drop rate and convergence ability than YOLOv7, thus proving that the improvement of the boundary regression loss function improves the convergence ability of the network.

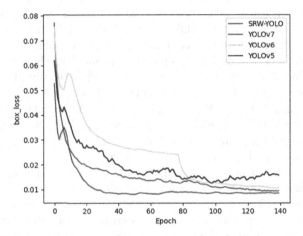

Fig. 9. Comparison of training box_loss curves of different models.

Table 3 lists the comparison results of the evaluation metrics of different models. ResNeXt-50 (32 × 4d), MobilenetV3-YOLOv4 and Ghost-YOLOv4 are Zu's research methods. Compared with the mainstream YOLO series algorithms, the performance of these methods in small target Lentinula Edodes logs contamination detection needs to be improved. Compared to other models, although the detection speed of the SRW-YOLO model proposed in this study is not the highest, it is much better than other models in the evaluation metrics of mAP@0.5, Recall, and mAP@0.5:0.9, This allows the model to maintain a good balance between detection accuracy and real-time performance.

Table 3. Comparison of evaluation indicators of different models.

Model	mAP@0.5	Recall	mAP@0.5:0.9	FPS
ResNeXt-50 (32 × 4d)	93.32%	91.12%	87.13%	17
MobilenetV3-YOLOv4	92.16%	89.11%	84.77%	30
Ghost-YOLOv4	93.04%	91.15%	86.56%	36
YOLOv5l	94.79%	92.68%	87.51%	67
YOLOv6m	95.82%	93.29%	91.32%	48
YOLOv7	96.31%	92.80%	91.66%	56
SRW-YOLO	98.62%	96.43%	92.39%	62

At the same time, in order to further demonstrate the superiority of the SRW-YOLO model improvement strategy, Table 4 lists the comparison results of the evaluation indicators of YOLOv7 and SRW-YOLO in three classes of Lentinula Edodes logs contamination detection. Compared with YOLOv7, SRW-YOLO has improved to varying degrees in the evaluation indicators of Precision, Recall and mAP@0.5. Among them, the original model has the worst effect in detecting Aspergillus flavus contaminated Lentinula Edodes logs, but the SRW-YOLO model improves Precision, Recall and mAP@0.5 by 8.12%, 5.33% and 2.36% respectively compared with YOLOv7. This shows that the SRW-YOLO model proposed in this article has more advantages in actual detection and can accurately detect different classes of Lentinula Edodes logs.

Table 4. Comparison of evaluation indicators of different classes.

class	model					
	YOLOv7			SRW-YOLO		
	Precision	Recall	mAP@0.5	Precision	Recall	mAP@0.5
Normal	95.62%	94.44%	99.53%	97.45%	96.81%	99.71%
Aspergillus flavus	87.71%	93.32%	95.37%	95.83%	98.65%	97.73%
Trichoderma viride	95.75%	86.16%	93.22%	99.73%	91.24%	97.31%

For a more intuitive understanding of the performance of the models, Fig. 10 shows the detection results of the four models for a randomly selected image in the test set, with the red box in the figure showing the area contaminated by Trichoderma viride. Although all four models are able to detect the type of Lentinula Edodes logs, YOLOv5l, YOLOv6m, and YOLOv7 have lower confidence in the detection of the target and poorer detection results. In contrast, SRW-YOLO has obvious superiority with 95% target confidence and accurately detects small object contaminated areas.

In summary, the Lentinula Edodes log contamination detection model proposed in this study has strong generalization ability and robustness. During the Lentinula Edodes logs cultivation stage, this model can better locate areas with small contamination objects in Lentinula Edodes logs and accurately detect the type of Lentinula Edodes log contamination.

Original picture

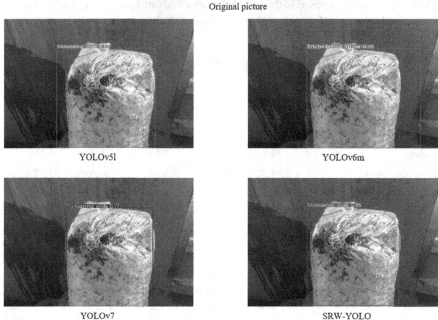

YOLOv5l   YOLOv6m

YOLOv7   SRW-YOLO

Fig. 10. Detection results of different models.

4   Conclusion

In this study a model for small object Lentinula Edodes log contamination detection (SRW-YOLO) was constructed on the basis of YOLOv7. SPD-Conv was introduced in the MP module of the feature fusion network to improve the learning ability of the model for small object contamination location and semantic information of Lentinula Edodes logs; the ELAN structure in the backbone network was reparameterized and the RepVGG architecture was used to realize the decoupling of training and inference to efficiently and accurately detect the types of Lentinula Edodes logs; the WIoU loss function was set as the boundary regression loss of the network function to reduce the competitiveness of high-quality anchor boxes while minimizing harmful gradients generated by low-quality samples to improve the overall performance of Lentinula Edodes log contamination condition detection. To verify the effectiveness of the model, ablation experiments and comparison experiments were conducted on a dataset. The experimental results showed that the detection accuracy of SRW-YOLO reached 97.63%, which was 4.62% better than that of YOLOv7. Compared to the current mainstream one-stage object detection model, the detection of small object Lentinula Edodes log contamination by SRW-YOLO is significantly better.

However, there are still some areas for improvement. The current Lentinula Edodes log dataset has a relatively simple background, and the model may not perform well when the background is more complex or the data collection environment is dimmer. Therefore, in subsequent work, the dataset will be further improved and the proposed Lentinula Edodes log contamination detection method will be optimized.

References

1. Cao, Z., Wang, S., Zheng, S., et al.: Identification of Paecilomyces variotii and its interaction with Lentinula edodes mycelium. North Horticulture **509**(14), 116–125 (2022)
2. Wang, Y., Liu, Z., Feng, Y., et al.: Study on the infection process of Trichoderma in the production of Lentinus edodes. Seed **40**(6), 131–141 (2021)
3. Yao, Q., Gong, Z., Si, H., et al.: Study on the formulation of culture substrate of lentinus edodes with resistance to hybrid bacteria. Chin. J. Edible Fungi **39**(10), 56–58 (2020)
4. LeCun, Y., Bengio, Y., Hinton, G.: Deep learning. Nature **521**(7553), 436–444 (2015)
5. O'Mahony, N., Campbell, S., Carvalho, A., Harapanahalli, S., Hernandez, G.V., Krpalkova, L., Riordan, D., Walsh, J.: Deep learning vs. traditional computer vision. In: Arai, K., Kapoor, S. (eds.) CVC 2019. AISC, vol. 943, pp. 128–144. Springer, Cham (2020). https://doi.org/10.1007/978-3-030-17795-9_10
6. Kim, J., Kim, B., Roy, P., et al.: Efficient facial expression recognition algorithm based on hierarchical deep neural network structure. IEEE Access **7**, 41273–41285 (2019)
7. Jeppesen, J., Jacobsen, R., Inceoglu, F., et al.: A cloud detection algorithm for satellite imagery based on deep learning. Remote Sens. Environ. **229**, 247–259 (2019)
8. Liu, J., Wang, X.: Plant diseases and pests detection based on deep learning: a review. Plant Methods **17**(22), 1–18 (2021)
9. Si, M., Deng, M., Han, Y.: Using deep learning for soybean pest and disease classification in farmland. J. Northeast. Agric. Univ. **26**(1), 64–72 (2019)
10. Jiang, P., Chen, Y., Liu, B., et al.: Real-time detection of apple leaf diseases using deep learning approach based on improved convolutional neural networks. IEEE Access **7**, 59069–59080 (2019)
11. Zu, D., Zhang, F., Wu, Q., et al.: Disease identification of Lentinus edodes sticks based on deep learning model. Complexity **2022**, 1–9 (2022)
12. Zu, D., Zhang, F., Wu, Q., et al.: Sundry bacteria contamination identification of Lentinula Edodes logs based on deep learning model. Agronomy **12**(9), 2121 (2022)
13. Han, J., Ding, J., Xue, N., et al.: ReDet: a rotation-equivariant detector for aerial object detection. In: Proceedings of the IEEE/CVF Conference on Computer Vision and Pattern Recognition, pp. 2786–2795. IEEE, Nashville, TN, USA (2021)
14. Zand, M., Etemad, A., Greenspan, M.: Oriented bounding boxes for small and freely rotated objects. IEEE Trans. Geosci. Remote Sens. **60**, 1–15 (2021)
15. Yu, D., Xu, Q., Guo, H., et al.: Anchor-free arbitrary-oriented object detector using box boundary-aware vectors. IEEE J. Sel. Topics Appl. Earth Observations Remote Sens. **15**, 2535–2545 (2022)
16. Zhu, X., Lyu, S., Wang, X., et al.: TPH-YOLOv5: Improved YOLOv5 based on transformer prediction head for object detection on drone-captured scenarios. In: Proceedings of the IEEE/CVF International Conference on Computer Vision, pp. 2778–2788. IEEE, Montreal, BC, Canada (2021)

17. Benjumea, A., Teeti, I,, Cuzzolin, F., et al.: YOLO-Z: Improving small object detection in YOLOv5 for autonomous vehicles (2021). arXiv preprint arXiv:2112.11798

18. Wang, C., Bochkovskiy, A., Liao, H.: YOLOv7: Trainable bag-of-freebies sets new state-of-the-art for real-time object detectors (2022). arXiv preprint arXiv:2207.02696

19. Sunkara, R., Luo, T.: No more strided convolutions or pooling: a new CNN building block for low-resolution images and small objects. Lect. Notes Comput. **13715**, 443–459 (2023)

20. Wang, C., Liao, H., Yeh, I.: Designing network design strategies through gradient path analysis (2022). arXiv preprint arXiv:2211.04800

21. Ding, X., Zhang, X., Ma, N., et al.: RepVGG: making VGG-style convents great again. In: Proceedings of the IEEE/CVF Conference on Computer Vision and Pattern Recognition, pp. 13733–13742. IEEE, Nashville, TN, USA (2021)

22. Tong, Z., Chen, Y., Xu, Z., et al.: Wise-IoU: Bounding Box Regression Loss with Dynamic Focusing Mechanism (2023). arXiv preprint arXiv:2301.10051

23. Selvaraju, R., Cogswell, M., Das, A., et al.: Grad-CAM: visual explanations from deep networks via gradient-based localization. In: Proceedings of the IEEE International Conference on Computer Vision, pp. 618–626. IEEE, Venice, Italy (2017)

A Study of Sketch Drawing Process Comparation with Different Painting Experience via Eye Movements Analysis

Jun Wang[1], Kiminori Sato[2], and Bo Wu[2(✉)]

[1] Graduate School of Bionics, Computer and Media Sciences, Tokyo University of Technology,
Hachioji 192-0982, Tokyo, Japan
g2123002c8@edu.teu.ac.jp
[2] School of Computer Science, Tokyo University of Technology,
Hachioji 192-0982, Tokyo, Japan
{satohkmn,wubo}@stf.teu.ac.jp

Abstract. When in a situation where there is a language barrier, facilitate the exchange through sketching is an efficient solution if a person wishes to keep the conversation going. However, not everyone has sketch basis, the method of using sketch to facilitate the exchange requires not only the foundation of sketch, but also the ability to sketch in a short period of time. Therefore, according to design and apply a set of experiments, we focus on analyzing the eye movement data of the subjects in the sketch of imaginary object shapes and compare the differences in the sketch between the experienced painters and the novice. Specifically, we invited 16 subjects to participate in sketching (e.g., a watch) on the canvas, and their eye movement data was collected by a glasses eye tracker while sketching. The results of analysis by Mann-Whitney U test showed that the novice's gaze was skewed to the left and down as a whole, the gaze was scattered, and the fixation on the central position of the picture was not focused and sustained. In addition, experienced painters put the sketch content in the center of the picture to attract the attention of the viewer and effectively convey the important information of the picture. The research results combine technology with art, which can help novice to quickly improve the efficiency and quality of sketch and carry out a wider range of art research and application.

Keywords: Eye movements analysis · Gaze detection · Expert-novice differences · Drawing skills · Experiment design

1 Introduction

People often experience language barriers when travelling abroad [1]. In order to keep the conversation going, a proven and effective method is using gestures or sketching as auxiliary method to promote communicate while speaking [2]. However, gestures vary from country to country, and even numbers such as 6, 7, 8, and 9 vary greatly [3]. By taking a specific scenario as an example, a tourist who lost his wallet sought assistance

H. Jin et al. (Eds.): GPC 2023, LNCS 14503, pp. 47–60, 2024.
https://doi.org/10.1007/978-981-99-9893-7_4

from the police. Tourists sketch the appearance of the wallet, allowing the police to better identify the characteristics of the lost wallet. Therefore, communicating through sketching is an effective solution [2].

Unfortunately, not all people have the basis of sketching, and it is difficult for people who do not have the basis of sketching or who sketch slowly to communicate with this method. Because they have not learned drawing knowledge, they often do not know where to start drawing, and they may be uncertain or afraid to draw, etc., so it is difficult for them to complete the sketch in a short time, and even the specific content of the sketch drawn by some people who have no sketching foundation is difficult to recognize.

As a basic art form and skill, sketch is the foundation and important part of visual art [4]. When learning and practicing sketch, many people often face problems such as difficulty in grasping the structure, perspective and proportion of the body, and even frequently modify, wipe and redraw in the process of drawing. In addition to individual factors, these problems are also related to the lack of basic education, imperfect teaching methods and other factors.

On the other hand, the use of eye tracking in social sciences, including education, has increased significantly [5]. The technique is now routinely used to investigate the learning process as well as to provide guidance on how to facilitate it. However, although eye-tracking has been widely used in education, there are few examples of its application in fine arts education.

As a common eye tracking technology in recent years, eye tracker can record and track the dynamic changes of human eyes in the rendering process in real time [6], and can accurately reflect the human cognition, perception and thinking process from many aspects such as details, movement and action [7]. Therefore, the use of eye tracker technology combined with painting software to analyze the eye movement behavior and painting process of learners with different levels can not only find the problems and difficulties of learners in the process of painting, but also analyze the sketching skills of experienced painters, which can provide reference for people who have never learned the basis of sketching.

Different from traditional eye-tracking devices, Tobii Pro Glasses 3 can collect high-precision eye-movement data while the user is painting by placing multiple sets of small cameras on the lens [8]. The results will contribute to a better understanding of human visual processing, improve the efficiency and quality of sketch drawing, and provide help for a wider range of art research and application.

In this study, we recorded the eye movements of experienced painters and novice in drawing sketches using Tobii Pro Glasses 3 eye tracker. By Using the collected eye movements data, the Mann-Whitney U statistical test was used to analyze the differences in "preferred viewing position" and "painting composition method" when drawing between experienced and novice scrolls. Based on this, we also generate heat map and scan path of the emphasis and sequence of experienced painters and novice during painting in the software to confirm our analysis results.

The contribution of this paper are: (1) By analyzing the sketching process of experienced painters, a proven method of drawing and composition was identified. (2) help novice understand their shortcomings, and learn the basic visual observation, painting skills and composition methods of sketch. (3) The research providing a new analytical approach, which the results can provide help for the related art research and application development.

2   Related Works

In this session, we will look at the application of eye trackers to art-related research questions and research demonstrating that sketch writing can effectively communicate. We will cover the application of Eye Tracker to art-related research problems, research on the differences between professional and non-professional studies, and research that demonstrates that sketch can effectively communicate.

2.1   Research on the Application of Eye Tracker in Art

The use of eye tracker for eye-tracking in the social sciences, including education, has increased significantly [5]. The technique is now routinely used to investigate the learning process as well as to provide guidance on how to facilitate it. Tseng PH et al. track that human eye fixations are biased toward the center of natural scene stimuli, and this bias affects the evaluation of computational models of attention and eye-dynamic behavior [9]. Simone F et al. revealed differences in visual attention between painting professionals and non-professionals during mental imagination by using eye tracker technology. This helps to further explore the relationship between painting professionalism and visual attention, and to conduct in-depth research on cognitive mechanisms in the process of painting and creation [10]. These findings have important implications for our understanding of art, and eye tracker technology provides us with objective data to help us interpret the way participants allocate attention and process information in drawing.

2.2   Studies Using Eye Trackers to Analyze Differences Between Professional and Non-Professional Subjects

Stine V et al. in their study comparing the eye movement patterns of artists and participants with no artistic training, artists showed a preference for scanning structural and abstract features, while untrained participants focused more on human features and objects. Artists also showed changes in viewing strategies during the memory task, with increased attention to object and person features. In addition, artists had better memory for image features compared to untrained participants [11]. Hannah F C et al. investigated cultural differences in eye movement patterns between American and Chinese participants when viewing photographs. Americans paid more attention to focal objects, while Chinese participants made more saccades of the background, suggesting that cultural differences in cognitive processing may stem from different attention to scene elements [12]. Kevin A P et al. examined differences in brain activity between non-specialists and specialists when observing eye shifts. Using functional magnetic resonance imaging (FMRI) and eye tracker monitoring, it was found that professionals showed large changes in blood oxygen levels when observing eye movement in brain regions such as supratemporal gyrus, and the delay of observation time and the correctness of observation target also had significant effects on brain activity [13]. These findings highlight differences in eye movement patterns and brain activity between professional and non-professional populations. Artists tend to focus on structural and abstract features and show better performance on memory tasks. Cultural differences may have an impact on eye movement patterns, resulting in different levels of attention to different scene elements. Professionals showed greater brain activity in response to eye movement. These findings contribute to a deeper understanding of the differences in cognitive and visual processing between professional and non-professional populations.

2.3 Cross-Cultural Communication Using Sketch Can Effectively Communicate the Study

As a communication tool in the application of collaborative design, sketch in the case of language barrier, more clearly express and understand the design intention. Y Wu et al. designed to explore the role of sketching in communication and communication by observing and analyzing the use of sketching by team members in the collaborative process [14]. Christine A believes that painting is a powerful means of communication that allows people to express themselves effectively and convey information [15]. Jang A et al. introduced an online interface called BodyDiagrams, which combined drawing and text to enable patients to graphically express pain symptoms and enhance the ability of text description. And through the evaluation of medical professionals, it was found that patients were more confident that their diagrams could be interpreted correctly, medical professionals considered the diagrams to be more informative than textual descriptions, and both groups indicated that they would be more inclined to use diagrams to communicate physical symptoms in the future [16]. These findings show that using sketches as a communication tool can help people more clearly express and understand design intentions or physical symptoms.

3 Methodology

This section details the hardware used, the variable measurements, and experimental design.

As shown in Fig. 1, Tobii Pro Glasses 3 is a lightweight wearable eye tracker technology suitable for capturing and analyzing natural visual behaviors in real-world environments [17]. It can accurately record what a person is focusing on during free movement in the real world, providing objective data for human visual gaze. In addition, the device is equipped with a front-facing scene camera that can provide a wider field of view (106° horizontal, 95° vertical) and capture video data [18]. These two types of data can be mixed into a special video that dynamically displays the user's focus area in the field of view using red dots.

3.1 Hardware

Eye tracking is a non-invasive behavioral measure that provides information about the processes in which the painters is involved at each moment of the sketch [19]. During

Fig. 1. Tobii Pro Glasses 3 Eye Tracker

Table 1. Details of eye-movement measures

Measures	Detail
TFD	The total length of time the subject's gaze remained on a specific area or target
FC	The total number of looks performed by the subjects during the experiment
TVD	Refers to the total amount of time spent visiting or staying in a specific location during a specific time
VC	The total number of times a particular location was visited during a particular time

sketching, the eyes were fixed together as they gathered information in saccades and moved between dots [20, 21]. Eye tracking allows researchers to objectively track and record what subjects are doing during comprehension tasks, for how long they are doing it, and in what order. For example, as shown in Table 1, the data of Total Fixation dur (TFD), Fixation count (FC), Total Visit duration (TVD) and Visit count (VC) can be collected for analyzing.

TFD: The total gaze duration is obtained by summing the duration of all fixations in a specific area in milliseconds. FC: measure from start to observe specific areas to stare at the end of the time for the first time, in milliseconds. TVD: The total viewing time is obtained by summing the durations of all fixations in a specific area in milliseconds. VC: This counts the number of times a subject looks at a particular area.

3.2 System Design and Experimental Process

In the experiment, we invite our subjects wearing the eye tracker, and ask them to sketch on the canvas (drawing board on screen) with a specific proposition (e.g., a wallet) by imagining.

Through the collection and necessary preprocessing of eye tracking data, it is used for the next analysis.

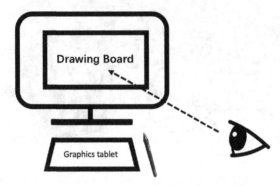

Fig. 2. System diagram design.

As shown in Fig. 2, in the experiment a digital tablet is used to control the paintbrush and paint on the canvas. All subjects do not allow to look at the tablet during the drawing process, all eye movements need to stay on the canvas.

As shown in Table 2, for comparison, three items with different shapes are selected as propositions, namely wallet (rectangle), headset (irregular shape) and wristwatch (circle). These three objects were chosen on the grounds that they are universally visible in everyday life, easy to lose, and relatively easy to draw for novice painters.

Table 2. The propositions needed in the experiment.

Proposition	Shape	Reasons for choice
Wallet	rectangle	1: Common in daily life
Headset	Irregular shape	2: Easy to lose 3: Different shapes
Wristwatch	Circle	4: Easy to draw relative to novice

Then, as shown in Fig. 3, all subjects were asked to wear eye tracker devices, and the experiment starts after the subject position their sight directly at the point right above the center of the computer screen. All subjects were asked to look at the point for two to three seconds before the subjects began to draw, which was conducive to data unification. At the end of experiment, we asked an artist with more than ten years of experience to rate all the images on a scale from a minimum of 1 to a maximum of 10.

3.3 Area of Interest (AOI) Setting

Area of Interest (AOI) setting is the selection of a displayed stimulus target area and the extraction of some eye movement indicators specifically in that area for statistical

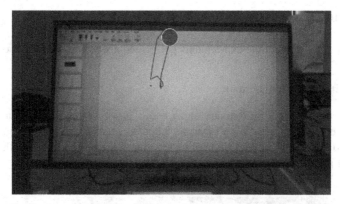

Fig. 3. Screenshot of eye tracker images of subjects during sketching.

analysis. According to the basic concepts of painting, composition and proportion are important for creating balanced, attractive images [22]. Therefore, As shown in Fig. 4, based on the basic concept of painting, we set five AOI to mark the areas that the painters look at. Divide the canvas into "Center", "Up", "bottom", "Left", "and Right". In Fig. 4, the left picture shows the screenshot interface of the actual AOI setting, and the right picture is the drawing of the AOI to see clearly.1

Fig. 4. Set AOI on the screen.

For each AOI, a total of four types of eye movement metrics were obtained through the eye tracker device. It includes Total Fixation dur (TFD), Fixation count (FC), Total Visit duration (TVD) and Visit count (VC). To be specific, TFD can calculate the total fixation time of the experimenter in each attention area and can obtain the index of the concentration and persistence of the experimenter's attention in the set AOI. FC indicates how many times the experimenter looks in the AOI. TVD is the sum of the time that the experimenter stayed in a particular area of interest. VC reflects the frequency of the experimenter's visits to the AOI, which increases whenever the experimenter looks at the AOI, and can know the allocation of attention and the degree of care for a particular domain.

4 Experiments, Analysisi and Discussion

In this section, we present the experiments performed according to the above design.

4.1 Experimental Environment

We conducted a series of preliminary experiments in a research room without natural light to make sure the light doesn't irritate subjects' eyes. As shown in Fig. 5, we set the canvas (display) and subject's line of sight to the same height and fixed the distance from the subject to the digital tablet and the display.

Fig. 5. Experimental environment.

4.2 Subjects

As shown in Table 3, we invited 16 subjects to our experiment, include 10 with no sketching-related experience, and 6 with more than one year of sketching-related knowledge. Before the experiment, the subjects were trained on how to wear the eye tracker and the use of the digital tablet during the experiment (e.g., instructions for pens and erasers).

4.3 Analysis Results and Discussion

To identify differences in eye movements during sketching between those with and without sketching experience, the Mann-Whitney u test was used. In this section, we

Table 3. Population Statistics.

Subject	Nationality	Experience	Subject	Nationality	Experience
1	China	novice	9	Japan	novice
2	China	novice	10	China	experienced
3	China	novice	11	China	experienced
4	Japan	novice	12	China	experienced
5	Japan	novice	13	China	experienced
6	Japan	novice	14	China	experienced
7	China	novice	15	China	novice
8	Japan	experienced	16	China	novice

compare the painters eye movement metrics for each AOI separately, including Center, Up, Under, Left, and Right. The relevant analyses were performed using SPSS 29.

Before statistical analysis, we preprocessed the data, and we checked the completeness and usability of each subject's picture.

The results on Table 3 show that there are significant differences between painters with different experience for almost all metrics of the two AOI we set, "Center" and "Under". It shows that experienced painters pay more attention to the "Center" and "Under" in composition, while novice may be more interested in other areas of the picture. The center of the picture is one of the most eye-catching positions visually. For experienced painters who have accumulated richer experience and knowledge in the process of learning to draw, by placing important elements in the center of the picture to attract the attention of the viewer, it can effectively convey the important information of the picture. In contrast, novice may lack an understanding of composition and visual organization, and thus exhibit more exploration and randomness in composition, possibly spreading their attention over the whole picture and exploring elements from different domains.

Moreover, since we have the data of score for all the images on a scale from a minimum of 1 to a maximum of 10, according to the professional perspective from the picture composition ratio and can intuitively recognize what is expressed by the picture to evaluate. According to the results, experienced painters sketch painters were significantly higher in painting proportion composition and picture expression than novice (Sig. <0.001) (Table 4).

For all results which mentioned above, similar inferences can be re-confirmed by the Heat map and Scan path. The Heat map represents the hotspot distribution of the painter's fixations, while the Scan path shows the duration and number of their fixations.

In the Heat map of experienced painters and novice shown in Fig. 6, we observe that the hot spots are clearly concentrated in the central position of the picture, indicating that experienced painters tend to gaze and focus on the elements in the center of the picture. It also reflects the in-depth understanding and application of the principles of composition by experienced painters.

Table 4. The results of SPSS analysis are summarized.

Measure	Drawing experience	N	Mean Rank	U	W	Z	Sig	Effect size
Center_TFD*	novice	31	20.45	138.000	634.000	2.705	0.007	0.494
	experienced	17	31.88					0.676
Left_TFD	novice	31	26.00	217.000	370.000	1.035	0.301	0.189
	experienced	17	21.76					0.259
Rirght_TFD	novice	31	23.53	233.500	729.500	0.850	0.395	0.155
	experienced	17	26.26					0.213
Under_TFD*	novice	31	21.58	173.000	669.000	2.068	0.039	0.378
	experienced	17	29.82					0.517
Up_TFD	novice	31	24.68	258.000	411.000	0.119	0.905	0.022
	experienced	17	24.18					0.030
Center_FC	novice	31	21.69	176.500	672.500	1.877	0.060	0.343
	experienced	17	29.62					0.469
Left_FC	novice	31	26.00	217.000	370.000	1.041	0.298	0.190
	experienced	17	21.76					0.260
Rirght_FC	novice	31	23.66	237.500	733.500	0.738	0.461	0.135
	experienced	17	26.03					0.185
Under_FC*	novice	31	25.26	160.000	656.000	2.372	0.018	0.433
	experienced	17	23.12					0.593
Up_FC	novice	31	25.26	240.000	393.000	0.509	0.611	0.093
	experienced	17	23.12					0.127
Center_TVD*	novice	31	20.29	133.000	629.000	2.813	0.005	0.514
	experienced	17	32.18					0.703
Left_TVD	novice	31	26.00	217.000	370.000	1.035	0.301	0.189
	experienced	17	21.76					0.259
Rirght_TVD	novice	31	23.60	235.500	731.500	0.794	0.427	0.145
	experienced	17	26.15					0.199
Under_TVD*	novice	31	21.55	172.000	668.000	2.091	0.037	0.382
	experienced	17	29.88					0.523
Up_TVD	novice	31	24.69	257.500	410.500	0.130	0.897	0.024
	experienced	17	24.15					0.033

(continued)

Table 4. (*continued*)

Measure	Drawing experience	N	Mean Rank	U	W	Z	Sig	Effect size
Centre_VC	novice	31	25.06	246.000	399.000	0.379	0.705	0.069
	experienced	17	23.47					0.095
Left_VC	novice	31	25.87	221.000	374.000	0.953	0.341	0.174
	experienced	17	22.00					0.238
Rirght_VC	novice	31	23.56	234.500	730.500	0.823	0.410	0.150
	experienced	17	26.21					0.206
Under_VC*	novice	31	21.29	164.000	660.000	2.286	0.022	0.417
	experienced	17	30.35					0.572
Up_VC	novice	31	25.24	240.500	393.500	0.499	0.618	0.091
	experienced	17	23.15					0.125
Score*	novice	31	16.00	0.000	496.000	5.790	< .001	0.026
	experienced	17	40.00					0.035

experienced   novice

Fig. 6. A comparison of Heat Map between experienced painters and novice

In contrast, novice have a more scattered gaze distribution. Indicates that the gaze in the center of the picture is less concentrated and sustained than that of experienced painters. The reason may be the lack of experience and knowledge in composition and the lack of attention and understanding of the central position of the picture.

This result is further supported by the Scan path of the novice and experienced painters shown in Fig. 7. The gaze map shows longer durations and more fixations in the central frame position for experienced painters, while the novice shows shorter durations and more fixations.

On the other hand, there is no significant difference in the index of "Left" AOI between experienced painters and novice. However, in the experiment, through observation, we can obviously feel that the novice tends to draw to the left, while the experienced painter generally concentrates their sketch content in the middle of the picture.

Therefore, we compared the images of experienced painters and novice after the experiment, as shown in Fig. 8. Through visual observation, we thought that there would be a significant gap between experienced painters and novice for the indicators on the left. On the contrary, there is no significant indicator for "Left" AOI As shown in Table 3.,

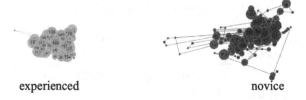

experienced   novice

Fig. 7. A comparison of Scan path between experienced and novice

the relevant indicators did not appear significant indicators about the "Left" AOI of experienced painters and novice.

Although a statistically significant difference was not reached, we still conducted an exploratory analysis and interpretation of this trend. In the experiment, we noticed that the novice showed more fixations and concentration on the left side of the canvas during drawing. This could be due to factors such as brain processing, hand habits or attentional bias. Such results may have been affected by the limitations of the sample size and experimental design. Factors such as the individual differences of painters in the experiment may have an impact on the results. In addition, the lack of sample size may also affect the judgment of significance.

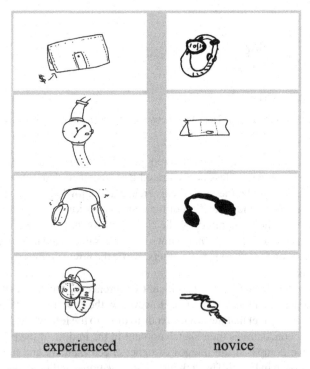

Fig. 8. A set of drawings of an experienced painters and novice

5 Conclusion

In order to compare the preference and delineation methods of experienced painters and novice painters when observing objects during sketching, a series of experiments and corresponding comparative analysis were designed and carried out using eye tracker.

Based on the 49 sets of eye movement data of 16 painters with different experience, the results of the Mann-Whitney u test show that there are significant differences between different experienced painters for the AOI of the center canvas and the upper canvas. In addition, we found a sketch artist with ten years of experience to score. The experienced painters sketch artist was significantly higher than the novice in painting proportion, composition and picture expression (Sig. <0.001). This result can provide statistical evidence of differences in composition between experienced painters and novice, demonstrating the importance of an artist's knowledge, skills, and aesthetic preferences for composition decisions.

Future research can explore the reasons for these differences and provide specific guidance and support in the field of education and training, which is expected to promote the improvement of novice in composition. In this experiment, we have only one teacher who is responsible for distributing the image scores. There may be concerns about potential bias or subjectivity in the evaluation process. In the future, we plan to have multiple evaluators or adopt a standardized scoring system. It may also be considered in the future whether sketches can be recognized as an additional evaluation criterion in the presence of language barriers. The inclusion of these aspects will broaden the scope of the study.

Based on the experimental results, we observed a clear tendency for the novice to lean to the left during sketching, but the significance level did not reach the statistical significance of 0.05. Although the 0.05 level was not reached, we speculate that this reason may be related to the lack of familiarity with the basics of sketch. Novice lacks attention to the center during sketching, resulting in left gaze and concentration. In the future, we need further research to confirm this speculation.

Acknowledgments. Appreciations for all the workers who participated in the experiments. This work was supported by JSPS KAKENHI Grant Number 21K11876.

References

1. Liang, W., Zhou, X., Huang, S., Hu, C., Xu, X., Jin, Q.: Modeling of cross-disciplinary collaboration for potential field discovery and recommendation based on scholarly big data. Future Gener. Comput. Syst. **87**, 591–600 (2018)
2. Suwa, M., Tversky, B.: What do architects and students perceive in their design sketches? a protocol analysis. Des. Stud. **18**(4), 385–403 (1997)
3. Kendon, A.: Gesture: Visible Action as Utterance. Cambridge University Press, pp. 123–125 (2004)
4. Kelly, A.E., Lesh, R.A., Baek, J.Y. (eds.): Handbook of Design Research Methods in Education: Innovations in Science, Technology, Engineering, and Mathematics Learning and Teaching, pp. 203–215 (2008)

5. Kaminskiene, L., Horlenko, K., Chu, L.Y.: Applying eye-tracking technology in the field of entrepreneurship education. In: Artificiality and Sustainability in Entrepreneurship, pp. 163–187 (2022)
6. Wu, B., Wu, Y., Nishimura, S., Jin, Q.: Analysis on the subdivision of skilled mowing movements on slopes. Sensors **22**(4), 1372 (2022)
7. Wu, B., Wu, Y., Dong, R., et al.: Behavioral analysis of mowing workers based on hilbert–huang transform: an auxiliary movement analysis of manual mowing on the slopes of terraced rice fields. Agriculture **13**(2), 489 (2023)
8. Zhou, X., Jin, Q.: A heuristic approach to discovering user correlations from organized social stream data. Multimed. Tools Appl. **76**(9), 11487–11507 (2017)
9. Tseng, P.-H., Carmi, R., Cameron, I.G.M., Munoz, D.P., Itti, L.: Quantifying center bias of observers in free viewing of dynamic natural scenes. J. Vision **9**(7), 4, 1–16 (2009)
10. Fagioli, S., Sibilla, F., Lupetti, M.L.: Visual attention during mental imagery differs according to drawing expertise: an eye-tracking study. Perception **48**(3), 303–318 (2019)
11. Vogt, S., Magnussen, S.: Expertise in pictorial perception: eye-movement patterns and visual memory in artists and laymen. Advance online publication, Perception (2007)
12. Chua, H.F., Boland, J.E., Nisbett, R.E.: Cultural variation in eye movements during scene perception. Proc. Natl. Acad. Sci. **102**(35), 12629–12633 (2005)
13. Pelphrey, K.A., Singerman, J.D., Allison, T., McCarthy, G.: Brain activation evoked by perception of gaze shifts: the influence of context. Neuropsychologia **41**(2), 156–170 (2003)
14. Wu, Y., Lin, C.: Sketching as a communication tool: a case study on collaborative design. Int. J. Des. **11**(1), 111–124 (2017)
15. Alford, C.: Drawing: the universal language of children. New Zealand J. Teachers' Work **12**(1), 45–62 (2015)
16. Jang, A., MacLean, D.L., Heer, J.: BodyDiagrams: improving communication of pain symptoms through drawing. In: Proceedings of the SIGCHI Conference on Human Factors in Computing Systems (CHI '14). Association for Computing Machinery, pp. 1153–1162 (2014)
17. Zhou, X., Li, S., Li, Z., Li, W.: Information diffusion across cyber-physical-social systems in smart city: a survey. Neurocomputing **444**, 203–213 (2021)
18. Wu, B., Zhu, Y., Yu, K., Nishimura, S., Jin, Q.: The Effect of Eye Movements and Culture on Product Color Selection. Human-centric Computing and Information Sciences, 2020, vol. 10, no. 48 (2020)
19. Wu, B., Tosa, K., Sato, K.: A Walk-through type authentication system design via gaze detection and color recognition. In: IEEE Intl Conf on Dependable, Autonomic and Secure Computing, Intl Conf on Pervasive Intelligence and Computing, Intl Conf on Cloud and Big Data Computing, Intl Conf on Cyber Science and Technology Congress, pp. 1–5 (2022)
20. Miklošíková, M., Malčík, M.: The use of eye tracking for the analysis of students' learning process when working with a concept map. In: International Conference on Emerging eLearning Technologies and Applications (ICETA), pp. 227–232 (2016)
21. Wu, B., Nishimura, S., Zhu, Y., Jin, Q.: Experiment design and analysis of cross-cultural variation in color preferences using eye-tracking. In: Proc. MUE2019 (The 13th International Conference on Multimedia and Ubiquitous Engineering) (2019)
22. Arnheim, R.: Art and Visual Perception: A Psychology of the Creative Eye. University of California Press, pp. 372–391 (1954)

Review of Deep Learning-Based Entity Alignment Methods

Dan Lu, Guoyu Han, Yingnan Zhao[✉], and Qilong Han

College of Computer Science and Technology, Harbin Engineering University, Harbin
150009, China
{ludan,zhaoyingnan}@hrbeu.edu.cn

Abstract. Entity alignment aims to discover different references to the
same entity in different graphs, and it is a key technique for solving
graph-related problems. It has developed into one of the important tasks
in knowledge graphs and has received extensive attention from scholars
in recent years. Through entity alignment, data from multiple isolated
knowledge graphs with different sources and modes can be summarized
and classified, forming a more information-rich knowledge base. In early
research on entity alignment, researchers first proposed a class of alignment methods based on knowledge representation learning and verified
that these methods have significant improvements over traditional methods. However, entity alignment still has many defects and challenges to
be addressed, such as a lack of scalability, differences in language and
relationship type definitions in knowledge graphs from different sources,
which make entity alignment difficult. There are also problems such as
the need to improve the quality of large-scale knowledge graph data
and optimize computational efficiency. Thus, it is difficult to perform
entity alignment on multiple knowledge graphs through simple translation and transformation. This paper discusses the deep learning-based
methods that have emerged in the field of entity alignment based on
the definition of entity alignment and data sets as standards, summa-
rizes the shortcomings and limitations of these methods, and introduces
commonly used data sets in the entity alignment task.

Keywords: Deep learning · Knowledge graph · Entity alignment ·
data set · Data quality

1 Introduction

With the development of internet technology and the continuous improvement
of high-tech computer technology, the amount of data within the internet is
also growing exponentially. More and more artificial intelligence technologies
are being widely applied in industry, from machine learning to deep learning.
Knowledge graph application is one of the technologies that is increasingly being
used in the development of science and technology.

H. Jin et al. (Eds.): GPC 2023, LNCS 14503, pp. 61–71, 2024.
https://doi.org/10.1007/978-981-99-9893-7_5

The concept of knowledge graph can be traced back to 2012 when it was introduced by Google with the main purpose of improving the search efficiency and quality of search results of its search engine. Google used semantic search to retrieve information from different websites, extracted key information from retrieved results, and stored them in a structured way for easy retrieval by users.

However, as the field of knowledge graph has rapidly developed in recent years, the diversity in the structure and construction standards of knowledge graphs has led to issues such as different structures, information redundancy, and diverse definitions. The fusion of multi-source and multi-modal knowledge graphs is the core of knowledge fusion and aims to form a unified knowledge identification and association for multiple knowledge graphs. Entity alignment is a key technique in the process of knowledge fusion, which aims to find equivalent entities among different knowledge graphs with different definitions and relationships. In knowledge graphs, entities are usually different objective things such as people, objects, events, and concepts that are clearly distinguishable from each other. The entity alignment method can help establish large-scale knowledge graphs with multiple sources and languages, but the different alignment methods used may lead to many defects in the alignment task. For example, in cross-lingual alignment tasks, the types and structures of data contained in different languages are also different. As shown in Fig. 1, the relationship between John and America corresponds to different structures in different graphs. Therefore, for cross-lingual alignment tasks, machine translation is often used to establish connections between entities, but low-quality machine translation may generate redundant data.

This paper focuses on introducing the research related to deep learning-based entity alignment methods, which are mainly composed of the following parts: related work, recent research on deep learning-based entity alignment problems in recent years, commonly used datasets in the entity alignment domain and the specific datasets adopted by each entity alignment method, conclusion, summarizing the content of this paper and further prospects for the future research and application directions of deep learning-based entity alignment methods.

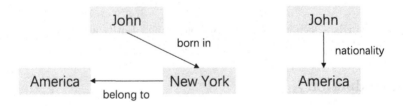

Fig. 1. Different structures around equivalent entities

2 Problem Definition and Common Datasets

2.1 Problem Definition

In knowledge graph, knowledge is stored in the form of triplets (h, r, t), where h represents the head entity, t represents the tail entity, and r represents the relationship between the head entity and the tail entity. The knowledge graph (KG) can be formalized as KG = (E, R, T), where E, R, and T represent the sets of entities, relationships, and triplets, respectively.

Assuming $KG_s = (E_s, R_s, T_s)$ and $KG_t = (E_t, R_t, T_t)$ are two knowledge graphs to be aligned, and $A = \{(e_{si}, e_{ti}) | e_{si} \in E_s, e_{ti} \in E_t\}_{i=1}^{m}$ represents the set of pre-aligned entity pairs, where e_{si} and e_{ti} represent two pre-aligned entities from the two knowledge graphs, and m represents the number of pre-aligned entity pairs. The goal of entity alignment task is to predict the remaining unknown aligned entity pairs $A' = \left\{ (e'_{si}, e'_{ti}) | e'_{si} \in E_{si}, e'_{ti} \in E_t \right\}_{i=1}^{n}$ [1]. As shown in Fig. 2, the purpose of entity alignment task is to discover equivalent entities existing in two knowledge graphs.

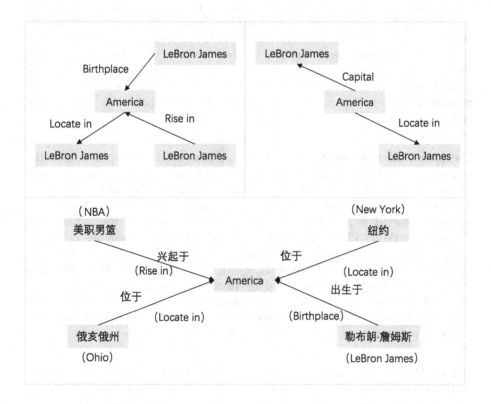

Fig. 2. Schematic diagram of entity alignment task

2.2   Common Datasets

For traditional entity alignment tasks, the main focus is on ontology alignment. Therefore, the performance evaluation of traditional entity alignment methods is mainly based on the evaluation system provided by the Ontology Alignment Evaluation Initiative (OVEI) [2]. In addition, some datasets are constructed from data generated by the Internet. SaraWagi et al. (2002) [3] obtained data from CiteSeer and the Local Telephone Company of Pune in India, with the datasets being Bibliography data and Address data, respectively. These two datasets contain a relatively small number of entities, with 254 and 300 entities in the Bibliography data and Address data, respectively. Cohen et al. (2002) [4] used the Cora, OrgName, Restaurant, and Parks datasets, which were sourced from the Cora project, organization names, restaurant guides, and national park name lists, respectively. The data in Cora records the bibliographic information of scientific papers, with entity attributes including author information and title. Jean-Mary et al. (2009) [5] used the OAEI (2008) dataset sourced from the Ontology Alignment Evaluation Initiative (OAEI) 2008. Arasu et al. (2010) [6] used the QRG and PUB datasets for entity alignment tasks, with the number of entities in each dataset being recorded as 2×10^6 (records) and 1.1×10^6. Suchank et al. (2011) [7] used the YAGO, DBpedia, and IMDB datasets, each containing a large number of entities, with 2795289, 2365777, and 4842323 entities, respectively. Lacosta et al. (2013) [8] used the IMDB and Freebase datasets, and Song et al. (2016) [9] used the triple dataset RKB and SWAT, with the number of entities in each dataset being recorded as 8.2×10^7 and 2.6×10^7. In addition to the datasets mentioned above, there are also datasets based on existing knowledge graphs and those constructed from Internet information.

Datasets Constructed Based on Existing Knowledge Graphs. Currently, many entity alignment datasets are constructed based on existing datasets such as YAGO, Wikidata, and DBpedia. DBpedia is one of the most widely used datasets in entity alignment research tasks, which is derived from structured content extracted from Wikipedia, and was first proposed by researchers from Leipzig University and Free University of Berlin. The first version was published in 2017, and as of 2016, there were six million entities in the DBpedia database, comprising nine and a half billion RDF triple data. The YAGO dataset is a comprehensive database constructed by researchers at the Max Planck Institute in Germany, which includes multilingual knowledge graphs from various data sources such as WordNet and GeoNames. It combines Wikipedia data with datasets from various sources, such as WordNet and GeoNames.

IMDB Data Set. IMDB was founded in 1990 as a film and television information database comprising of information about numerous movies, TV shows, and other related entities such as actors. Unlike databases such as YAGO and DBpedia which are constructed using text information extracted by algorithms with manual assistance, IMDB is built by individuals from around the world interested in contributing to the database. In some cases, the actors or creators recorded in the database are even invited to participate in its construction, which imparts a higher level of credibility to the IMDB database.

DBP15K. The origin of the DBP15K dataset can be traced back to 2015, when several scholars from Tsinghua University selected entries from the DBpedia dataset based on pre-defined rules. The selected entries were used to create a dataset comprising of data items in four different languages, namely Chinese, French, Japanese, and English. As such, DBP15K is often utilized in cross-lingual entity alignment tasks. Details of the DBP15K dataset can be seen in Table 1.

Table 1. DBK15K data set

Source	Data set		Number of entities	Number of relationships	Number of attributes	Number of relationship triples	Number of attribute triples
DBpedia	DBP15K	English	95680	2096	6066	233319	497230
		Japanese	65744	2043	5882	164373	354619
		English	105889	2209	6422	278590	576543
		French	66858	1379	4547	192191	528665
		English	98125	2317	7173	237674	567755
		Chinese	66469	2830	8113	153929	379684

3 Entity Alignment Based on Deep Learning

In traditional entity alignment research methods, feature matching methods based on similarity functions are primarily used, for example, using text similarity or edit distance to determine whether the source entity and the target entity are the same entity, or using the similarity of the source entity and the target entity's own structure in their respective graphs to make judgments, which require the calculation of similarity through the function settings. Traditional entity alignment methods often require manual definition of various entity features, which heavily depends on the personal experience of developers, leading to high time and manpower costs. In recent research on entity alignment, researchers have addressed these issues and achieved good performance by using knowledge embedding models and graph neural networks (GNNs) [10] to solve entity alignment tasks. Classic TransE-based [11] methods have been widely used, such as the MTransE [12] model. Researchers have adopted knowledge embedding models and graph neural networks (GNNs) to construct an entity network based on the relationships between entities and learn the embedding of entities. The GNN-Align model constructs a weighted network based on the relationships between entities and uses graph convolutional networks (GCNs) [13] to learn the embedding of entities. These methods project multi-source knowledge graph entities into a unified semantic vector space, and determine new aligned entities by calculating the vector distance between source entity nodes and target entity nodes in two knowledge graphs.

3.1   Cross Knowledge Graph Entity Alignment Based on Relationship Prediction

Hongren Huang et al. [14] proposed a novel entity alignment framework called RpAlign. They argue that traditional entity alignment methods face two major challenges: designing hyperparameters to balance embedding loss and alignment loss, and limited training data. The RpAlign model transforms the entity alignment task into a knowledge graph completion task, without the need for any additional alignment components. Compared to existing models that use entity vector distance, RpAlign defines a new relation called "anchor" alignment between entities, which predicts the relations between new aligned entities based on relation prediction. The RpAlign model framework mainly consists of three sub-modules: data expansion module, knowledge embedding module, and self-supervised learning module. The data expansion module mines more knowledge triplets from known entity pairs and relations to help the model learn more accurate knowledge representation. Data expansion techniques can fully exchange knowledge information between two knowledge bases, thus helping to supplement missing knowledge information in individual knowledge bases and assist the model to learn more accurate knowledge representation. RpAlign introduces the "anchor" relation as the relation r_α between aligned entities on two knowledge graphs, generating supervised triplets that can be added to the original training triplets, as shown in Equation (1).

$$T_\alpha = \{(e_s, \gamma^\alpha, e_t) \cup (e_t, \gamma_\alpha, e_s) \,|\, (e_s, e_t) \in A\} \qquad (1)$$

Let A be the set of aligned entity pairs, and e_s and e_t be the entity sets of the two knowledge graphs. Therefore, the number of triplets with the "anchor" relation in the training data of RpAlign is greater than the total number of entities on both knowledge graphs, thus alleviating the limited pre-aligned entity pair problem. These triplets enable the model to learn a scoring function for the "anchor" relation, which is used to measure the existence of aligned entities. After obtaining external triplets generated by the data expansion module in the RpAlign framework, the knowledge embedding module is used to learn embeddings of entities and relations in the mixed knowledge graph. PRotatE is chosen as the knowledge embedding model in RpAlign because it is more effective in modeling the symmetric/asymmetric relation patterns compared to other existing knowledge embedding models. PRotatE is a variant of the rotation model, where the model for entity embedding is determined by a constrained formulation $|h_i| = |t_i| = C$, where h_i and t_i denote head and tail entities, respectively. The knowledge embedding module uses negative sampling techniques to optimize the model, and the final loss function is shown in the equation.

$$L = -log\sigma\left(\gamma - d_r\left(h, t\right)\right) - \sum_{i=1}^{n} p\left(h_i', t_i' | h, r, t\right) log\left(d_e\left(h_i', t_i'\right) - \gamma\right) \qquad (2)$$

Here, σ is the sigmoid function, γ is the fixed margin, and $p\left(h_i', t_i' | h, r, t\right)$ represents the negative triplet distribution used in the knowledge embedding module. RpAlign adopts a self-supervised learning paradigm to address the limited training data problem. The concept of self-supervised learning is to annotate unlabeled data based on the predicted results after each training iteration, and then add the newly annotated data to the original training data in the next training iteration to retrain the model. RpAlign performs one-to-one entity alignment labeling for unaligned entities in both knowledge graphs. For each unaligned entity in the knowledge graph, RpAlign adds a virtual entity that is connected to all entities in the opposite knowledge graph through the "anchor" relation. The virtual entity is then aligned with the most similar entity in the opposite knowledge graph. During self-supervised training, RpAlign updates the score function by considering the "anchor" relation score and the negative triplet distances.

RpAlign finds the most likely aligned entities based on the value of the score function, and generates new aligned entity pairs according to Equation (3).

$$A_{new} = \left\{ \left(e_s, e_t | e_s \in E_s, e_t = max \oint_{r_a} (e_s, e_t) \right) \right\} \tag{3}$$

Due to the fact that the relations between two knowledge bases are not one-to-one, when the number of aligned entities is small, there is a chance that the triplets generated by the alignment relations in the data expansion module can be incorrect. The self-supervised learning mechanism can continuously generate new shared aligned entity pairs, improving the accuracy of the data expansion module on newly generated triplets. This means that the self-supervised module can reduce noise from the data expansion module. In summary, the self-supervised module of RpAlign can help the data expansion module to generate more cross-knowledge graph triplets. This method can effectively alleviate the noise problem caused by the lack of labeled data and improve the quality of the generated data by the data expansion module. RpAlign was evaluated for efficiency using two large-scale real datasets, DBP15K and DWY100K. The experimental results showed that RpAlign performed efficiently and achieved superior performance. Further research can explore how to reduce the computation time of the model and use more entity and relationship information to improve the conflict resolution mechanism of the model.

3.2 Semi Supervised Neighborhood Matching Model for Global Entity Alignment

Beibei Zhu et al. [15] found that most entity alignment models only consider the one-hop neighborhood information of candidate entity pairs and do not take into account the relationships with adjacent nodes. When two entities have the same neighborhood structure, the relationship is crucial in determining whether they can be aligned. Therefore, in addition to utilizing the structural information around entities, Beibei Zhu et al. also used relationship semantics to enhance entity alignment. Based on the premise that the larger the number of seed sets used for training, the better the model performance, Beibei Zhu et al. used a semi-supervised bi-directional nearest neighbor iterative strategy to expand the size of the training seed set, avoiding the need for manual labeling. In addition, to ensure the stability of entity alignment results, Beibei Zhu et al. performed global entity alignment from a comprehensive perspective and evaluated the performance of the SNGA model on three public cross-language datasets.

In their research, Beibei Zhu et al. iteratively expanded the seed sets. The number of seed sets between iterations was fixed, but in each subsequent iteration, new aligned entity pairs were generated, which were filtered and added to the seed set to achieve the goal of expanding the seed set. Using the expanded seed set as the input to the SNGA model improved the quality of embedding and neighborhood matching, which is beneficial for enhancing entity alignment.

Beibei Zhu et al. used GCNs to learn embedding representations, and the GCNs continuously optimized the model through backpropagation. They used an end-to-end approach to learn vector representations of nodes. Beibei Zhu et al. believed that non-end-to-end learning methods usually consist of different components, which may amplify errors between components due to separate training and have high complexity in combining different components. Compared with non-end-to-end learning methods, end-to-end learning methods can effectively reduce error accumulation and complexity.

Like most entity alignment methods, the SNGA model uses DBP15K as the dataset for training and testing. Its advantage is that it introduces relationship semantics into domain matching, combines domain matching, iterative strategy, and global entity alignment to effectively improve entity alignment rate, and learns entity representations during the iterative process by combining structural embedding and relationship embedding. The experimental results show that SNGA is better than many existing models, and sensitivity analysis also demonstrates the necessity and rationality of each part in the design of the SNGA model. However, there is still room for improvement in SNGA. The model only utilizes structural information and relationship semantics, but its semantics and string information are not fully utilized, so it is necessary to explore how to integrate other semantic information and string information to enhance the effect of entity alignment models in future work.

3.3   Entity Alignment Method Through Abstract and Attribute Embedding

Knowledge graph is a popular approach to store facts about real-world entities. However, its drawback is that the number of entities stored in different knowledge graphs is very limited. A embedding-based entity alignment method discovers equivalent relationships between entities by measuring the similarity between entity embeddings. Existing methods mostly focus on aligning entities based on their relationship structure and property information. However, when entities have few properties or relationship structures cannot capture meaningful representations of entities, this approach can be less efficient. A entity alignment method (EASAE) [16] based on abstract and property embeddings can solve this problem by using entity abstract information in the knowledge graph for entity abstract embeddings. EASAE uses bidirectional encoders from Transformers (BERT) to learn the semantics of entity abstracts, and learns entity representations by using relation triplets, attribute triplets, and abstracts.

For abstract embeddings, EASAE regards the text description associated with each entity as an abstract. For example, in DBPEDIA, most entities have a short abstract. Similarly, Wikidata also provides a summary text description of entities, which includes basic information about these entities. EASAE uses BERT to generate a set of word vectors from the abstract of a specific entity, to capture the main meaning of the entity. EASAE uses three different scoring functions to learn entity abstract embeddings, as shown in Equation (4).

$$\int_s = \|h_{sum} + \gamma - t_{sum}\|_{\frac{L1}{L2}}$$

$$\int_{st} = \|h_{sum} + \gamma - t\|_{\frac{L1}{L2}}$$

$$\int_{hs} = \|h + \gamma - t_{sum}\|_{\frac{L1}{L2}} \tag{4}$$

For the scoring function $\int_s$, h_{sum} and t_{sum} represent the representations of the head and tail entities respectively. In $\int_{st}$, h_{sum} captures the representation of the head entity using summary-based approach, while the tail entity uses relation-based embedding. However, in $\int_{st}$, this is completely reversed. In the equations, $\frac{L1}{L2}$ represents the $\frac{L1}{L2}$ norm. The entity embedding learning formula for the summary model is shown in Equation (5).

$$L_{SUM} = \int_s + \int_{st} + \int_{hs} \tag{5}$$

The loss function L_{SUM} is based on the summary-based entity representation, with the objective of minimizing l_{SUM}.

Entities are connected through relationships, which serve as features in knowledge graphs. In the relationship modeling part of EASAE, the structural relationships of triples are analyzed. TransE is used to interpret relationships as translation vectors from the head entity to the tail entity. For relationships, a similar approach as described in JSAE is adopted in EASAE, where, similar to Attribute, the weight α is used to control embedding learning of predicates on aligned triples. In order to learn structural embeddings, the EASAE model sets a target function l_{RE} as shown in Equation (6) for minimization.

$$L_{RE} = \sum_{(h,r,t) \in T_r} \sum_{h',r',t'} \in T'_r max\left(0, \gamma + \alpha\left(\int_r (h,t) - \int_r (h',t')\right)\right) \tag{6}$$

In this paragraph, α denotes $\alpha = \frac{count(r)}{|T|}$, γ represents the margin, $\int_r (h,t)$ and $\int_r (h',t')$ are the scoring functions for positive and negative triples respectively, T_r is the set of positive triples, T_r' is the set of corrupted triples, $\int_r$ is the likelihood function for relation triples, $count\,(r)$ is the occurrence count of relation r, and t represents the total number of triples in the merged knowledge graph. During the entity alignment process, EASAE combines the scores of three embedding models into an integrated method to achieve better predictive performance. It uses a weighted averaging technique to obtain the overall score of the model. In the EASAE model, instead of linearly combining the three embeddings, weights are assigned to each entity embedding to emphasize important components. According to the experimental results of the EASAE method, it has achieved better performance than baseline embedding methods, but more dataset validation is needed to determine the effectiveness of the model.

4 Conclusion

Knowledge graphs have become one of the successful applications of artificial intelligence in industrialization, where entity alignment plays a crucial role in achieving this success. This paper analyzes the entity alignment task starting from its basic concept, and discusses the research on existing entity alignment methods, their flaws and challenges. Then, recent deep learning-based entity alignment methods are briefly introduced.

This paper believes that the future direction for improving the entity alignment task is to more effectively utilize the known network structure and related information, by encoding the network structure information through multidimensional and deeper learning methods to improve task performance. In the future, further research can be conducted by modifying various experimental settings, such as altering the weights of relations and attributes, improving the embedding representation model, and pre-training on entity and attribute descriptions. In addition, large-scale and accurate entity alignment data are crucial for improving task performance.

Acknowledgement. This work was supported by the National Key R&D Program of China under Grant No. 2020YFB1710200.

References

1. Huang, H., Li, C., Peng, X., et al.: Cross-knowledge-graph entity alignment via relation prediction. Knowl.-Based Syst. **240**, 10781 (2022)
2. Euzenat, J., Ferrara, A., Hollink, L., et al.: Results of the ontology alignment evaluation initiative 2009 (2010)
3. Sarawagi, S., Bhamidipaty, A.: Interactive deduplication using active learning. In: Proceedings of the Eighth ACM SIGKDD International Conference on Knowledge Discovery and Data Mining, pp. 269–278 (2002)

4. Cohen, W.W., Richman, J.: Learning to match and cluster large high-dimensional data sets for data integration. In: Proceedings of the Eighth ACM SIGKDD International Conference on Knowledge Discovery and Data Mining, pp. 475–480 (2002)
5. Jean-Mary, Y.R., Shironoshita, E.P., Kabuka, M.R.: Ontology matching with semantic verification. J. Web Semant. **7**(3), 235–251 (2009)
6. Arasu, A., Götz, M., Kaushik, R.: On active learning of record matching packages. In: Proceedings of the 2010 ACM SIGMOD International Conference on Management of Data, pp. 783–794 (2010)
7. Suchanek, F.M., Abiteboul, S., Senellart, P.: PARIS: Probabilistic alignment of relations, instances, and schema (2011). arXiv preprint arXiv:1111.7164
8. Lacoste-Julien, S., Palla, K., Davies, A., et al.: SiGMA: simple greedy matching for aligning large knowledge bases. In: Proceedings of the 19th ACM SIGKDD International Conference on Knowledge Discovery and Data Mining, pp. 572–580 (2013)
9. Song, D., Luo, Y., Heflin, J.: Linking heterogeneous data in the semantic web using scalable and domain-independent candidate selection. IEEE Trans. Knowl. Data Eng. **29**(1), 143–156 (2016)
10. Scarselli, F., Tsoi, A.C., Gori, M., Hagenbuchner, M.: Graphical-based learning environments for pattern recognition. In: Fred, A., Caelli, T.M., Duin, R.P.W., Campilho, A.C., de Ridder, D. (eds.) SSPR /SPR 2004. LNCS, vol. 3138, pp. 42–56. Springer, Heidelberg (2004). https://doi.org/10.1007/978-3-540-27868-9_4
11. Bordes, A., Usunier, N., Garcia-Duran, A., et al.: Translating embeddings for modeling multi-relational data. In: Advances in Neural Information Processing Systems, vol. 26 (2013)
12. Chen, M., Tian, Y., Yang, M., et al.: Multilingual knowledge graph embeddings for cross-lingual knowledge alignment (2016). arXiv preprint arXiv:1611.03954
13. Kipf, T.N., Welling, M.: Semi-supervised classification with graph convolutional networks (2016). arXiv preprint arXiv:1609.02907
14. Huang, H., Li, C., Peng, X., et al.: Cross-knowledge-graph entity alignment via relation prediction. Knowl.-Based Syst. **240**, 107813 (2022)
15. Zhu, B., Bao, T., Wang, K., et al.: A semi-supervised neighborhood matching model for global entity alignment. Neural Comput. Appl. 1–21 (2023)
16. Munne, R.F., Ichise, R.: Entity alignment via summary and attribute embeddings. Logic J. IGPL **31**(2), 314–324 (2022)
17. Zhao, X., Zeng, W., Tang, J., et al.: An experimental study of state-of-the-art entity alignment approaches. IEEE Trans. Knowl. Data Eng. **34**(6), 2610–2625 (2020)
18. Sun, Z., Zhang, Q., Hu, W., et al.: A benchmarking study of embedding-based entity alignment for knowledge graphs (2020). arXiv preprint arXiv:2003.07743
19. Trisedya, B.D., Qi, J., Zhang, R.: Entity alignment between knowledge graphs using attribute embeddings. In: Proceedings of the AAAI Conference on Artificial Intelligence, vol. 33, no. 01, pp. 297–304 (2019)

VMD-AC-LSTM: An Accurate Prediction Method for Solar Irradiance

Jianwei Wang[1], Ke Yan[1,2], and Xiang Ma[1(✉)]

[1] Key Laboratory of Electromagnetic Wave Information Technology and Metrology of Zhejiang Province, College of Information Engineering, China Jiliang University, Hangzhou 310018, China
maxiang@cjlu.edu.cn
[2] Department of the Built Environment, College of Design and Engineering, National University of Singapore, Singapore 117566, Singapore

Abstract. Currently, solar power has become one of the most promising new power generation methods. But electricity cannot be stored directly and solar power has strong volatility, therefore the short-term accurate prediction of solar irradiance is of great significance to maintain the stable operation of the power grid. This work presents a novel decomposition integrated deep learning model, VMD-AC-BiLSTM, is proposed for ultra-short-term prediction of solar irradiance. The proposed model organically combines Variational Modal Decomposition (VMD), Multi-head Self-Attention Mechanism, One-Dimensional Convolutional Neural Network (1D-CNN) and Bidirectional Long and Short-Term Memory Network (BiLSTM). Firstly, the historical data are decomposed into several modal components by VMD, and these components are divided into stochastic and trend component sets according to their frequency ranges. Then the stochastic and periodicity of solar irradiance are predicted by two different prediction modules. The prediction results of the two modules are integrated at the end of the proposed model. Meanwhile, the proposed model also considers the complex effects of cloud type and solar zenith angle with stochasticity and periodicity in solar irradiance data, respectively. The experimental results show that the proposed model produces relatively accurate solar irradiance predictions under different evaluation criteria. And the proposed model has higher prediction accuracy and robustness compared to other deep learning models.

Keywords: Solar Irradiance · Deep Learning · Variational Modal Decomposition

1 Introduction

Electricity plays a very important role in the process of social and economic development [1]. So the amount of electricity generation is often used to measure the level of economic development of a country or region [2]. As of now, the most important source of electricity production is still the traditional thermal power generation. But the process of thermal power generation produces a large amount of carbon dioxide [3], which leads to the increasingly serious greenhouse effect. Therefore, there is an urgent need to transform the current power generation from traditional fossil energy sources to renewable energy sources [4]. It is expected that nearly 60% of the total power generation in 2040 will come from renewable energy sources, with wind and solar photovoltaic accounting for more than 50% [5, 6]. According to the International Renewable Energy Agency (IRENA), by the end of 2020, the global solar photovoltaic (PV) installed capacity has reached 714GW, and China's PV installed capacity has reached 253GW [7]. In short, PV power is one of the most promising new ways of generating electricity.

Unlike traditional thermal power generation, PV power generation is highly suscep-tible to the influence of external environmental factors [8]. As a result, it shows strong volatility and intermittency, which leads to difficulties in integrating PV power gen-eration methods into the power grid. Therefore, the power sector needs to accurately predict the PV power so that it can coordinate multiple generation methods. And PV power generation is closely related to solar irradiance [9], so solar irradiance data are often applied to PV forecasting.

Deep learning (DL) approaches have become widely used in time series prediction research in recent years, including studies of wind speed [10], electricity consumption [11] and photovoltaics [9]. The most popular is the Long Short-Term Memory Net-work (LSTM), a Recurrent Neural Network (RNN) version, whose special gating unit architecture overcomes gradient vanishing and explosion issues in RNNs and improves time series prediction [12]. The decomposition-integrated deep learning framework has been widely utilized for solar irradiance prediction, and has demonstrated noticeably better prediction results than a single deep learning model [13]. It is an efficient and reli-able method. The decomposition-integrated deep learning models used today for solar irradiance prediction, in our opinion, are still not sophisticated enough.

We visualized the solar irradiance data, taking the spring season as an example, as shown in Fig. 1. It is easy to find that solar irradiance has three characteristics: randomness, periodicity, and climatic trend. Among them, stochasticity which has no obvious regularity is the main reason why solar irradiance is difficult to predict accurately, so it needs a relatively more complex model with better data modeling ability to make predictions. It is well known that cloud motion is closely related to solar irradiance. If a certain cloud blocks out the sun, the light intensity at the ground level will drop dramatically, and this descent is accomplished in a very short period of time. Not only that, the pattern of the effect of different cloud types and weather on the light is different. So the proposed model takes into account the complex effect of the cloud type on the stochasticity in the data of solar irradiance. Unlike the randomness, the cyclic variation in the solar irradiance data is much simpler. It is determined by the sunrise and sunset due to the rotation of the earth, and this cyclic fluctuation has a strong regularity. It is analogous to the fluctuating form of the sine function when it is taken to its absolute value. The zenith angle of the sun accurately reflects this sinusoidal fluctuation, because they have the same period and a consistent tendency to change. At the same time, the amplitude of the sinusoidal function is gradually increasing, which we call the climatic trend. Unlike spring, winter shows a declining climate trend, and a relatively smooth climatic trend in the summer and fall. So we designed a low computational complexity cycle prediction module which takes into account solar zenith angle and climate trends for predicting gradually changing cyclic fluctuations.

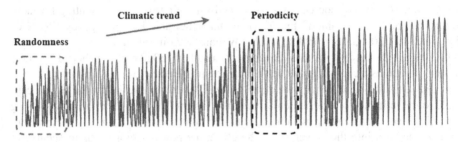

Fig. 1. Stochasticity, periodicity and climatic trends in spring solar irradiance.

Most existing decomposition-integrated deep learning methods input each sub-signal obtained from decomposition into their respective deep learning models separately [14, 15], ignoring the intrinsic correlation between different sub-signals. Moreover, most existing decomposition integration models use wavelet decomposition (WTD) or empirical modal decomposition (EMD) family of methods, but they have unavoidable drawbacks. Performing wavelet decomposition requires tedious basis function selection, but there are more than ten common wavelet basis functions, which have different properties. Each basis function also has a different filter length, and small changes in the basis function may lead to completely different predictions. Empirical modal decomposition suffers from mode aliasing and the number of decompositions cannot be specified. The Complete Ensemble Empirical Mode Decomposition with Adaptive Noise (CEEMDAN) and the Improved Complete Ensemble Empirical Mode Decomposition with Adaptive Noise (ICEEMDAN) solve the problem of mode aliasing to a certain extent, but their computational costs also become higher.

Based on the existing research problems and the characteristics of solar irradiance data, the main contributions of this paper are summarized as follows:

- Using VMD to decompose the historical solar irradiance data into several subseries with different frequencies, it overcomes the inherent defects of wavelet and empirical modal series decomposition. And the subseries obtained from the decomposition are divided into stochastic component set and trend component set, which are input to different prediction modules respectively.
- Two prediction modules of varying complexity were designed. The stochastic prediction module takes as input the set of stochastic components and considers the effect of cloud motion on the stochasticity of solar irradiance. The periodic prediction module takes into account the solar zenith angle and seasonal trends.
- Comparing our proposed VMD-AC-BiLSTM with common deep learning models and other comprehensive models for solar irradiance prediction. The experimental results show that the proposed model has higher prediction accuracy and robustness compared to other comparative models.

2 Related Works

With the rise of artificial intelligence technology, deep learning-based forecasting methods have become the preferred method for time series forecasting [16]. LSTM is the most popular single model approach for time series forecasting, which is an improved version of RNN, which solves the problems of gradient vanishing and gradient explosion of RNN [17]. Qing et al. proposed a new scheme for hour-by-hour solar irradiance forecasting based on LSTM utilizing weather forecast information, which takes into account the dependency between consecutive hours within the same day and substantially improves forecasting accuracy compared to machine learning models [18]. Recent studies have found that a hybrid model approach combining several single models can further improve prediction accuracy. Cascone et al. combined convolution with LSTM to design the ConvLSTM network to predict the power of household appliances in two phases, and achieved better prediction results than a single deep learning model [19]. Guo et al. combined a convolutional neural network (CNN) and Bidirectional Long Short-Term Memory Network (BiLSTM) for train travel time prediction through multi-feature fusion, which achieved higher prediction accuracy compared to other hybrid models [20]. Huang et al. designed a more complex model for hourly PV power prediction, which combined Conditional Generative Adversarial Network (CGAN), CNN, and BiLSTM, and based on historical data and random noise vectors to extract features through a complex model [7].

Many recent studies have combined deep learning with signal decomposition methods to constitute decomposition integration models. Compared with the conventional deep learning models, the decomposition integration model has stronger feature extraction ability, better generalization and robustness. Zeng et al. used the wavelet transform (WT) and nested short-term and long-term memory network (NLSTM) integration model for air quality prediction, decomposing the historical data sequence into several sub-sequences, inputting them into sub-models with the same structure respectively, and then integrating the output results of each sub-model to get the final prediction result [21]. In the field of solar irradiance prediction, Pi et al. put the different frequency sub-signals obtained after wavelet decomposition through more complex sub-prediction models containing CNN, LSTM, and attention mechanism respectively, and achieved good results [9]. Some other studies have considered the effect of meteorological parameters such as temperature on solar irradiance [22].

3 Methodology

3.1 Variational Modal Decomposition

Variational Modal Decomposition (VMD), proposed by Dragomiretskiy in 2013 [23], is a non-recursive signal decomposition method that reduces the non-smoothness of complex nonlinear sequences.

It is assumed that the measured signal $f(t)$ can be adaptively decomposed into multiple modal components u_k with center frequency ω_k and finite bandwidth, while the sum of estimated bandwidths of each mode is minimum. The constraint is that the sum of all modes is equal to the original signal, which can be expressed as:

$$\begin{cases} \min\limits_{\{u_k\}\{\omega_k\}} \left\{ \sum_k \left\| \partial_t \left[\left(\delta(t) + \frac{j}{\pi t} \right) * u_k(t) \right] e^{-j\omega_k t} \right\|_2^2 \right\} \\ \text{s.t.} \sum_k u_k(t) = f(t) \end{cases} \tag{1}$$

where $\delta(t)$ denotes the unit impulse function and $*$ denotes the convolution symbol. The above equation constrained optimization problem can be equated to an unconstrained optimization problem by augmenting the generalized Lagrangian function as follows:

$$L(\{u_k\}, \{\omega_k\}, \lambda) = \alpha \sum_k \left\| \partial_t \left[\left(\delta(t) + \frac{j}{\pi t} \right) * u_k(t) \right] e^{-j\omega_k t} \right\|_2^2$$

$$+ \left\| f(t) - \sum_k u_k(t) \right\|_2^2 + \langle \lambda(t), f(t) - \sum_k u_k(t) \rangle \tag{2}$$

where λ denotes the Lagrange multiplier and α denotes the penalty parameter. Initialize the parameters u_1, ω_1, λ and n, with the initial value of n set to 0. Set up the loop process so that $n = n + 1$, u_k, ω_k and λ are updated according to Eqs. (3–5):

$$u_k^{n+1} = \frac{f(\omega) - \sum_{i \neq k} u_i(\omega) + \frac{\lambda(\omega)}{2}}{1 + 2\alpha(\omega - \omega_k)^2} \tag{3}$$

$$\omega_k^{n+1} = \frac{\int_0^\infty \omega |u_k(\omega)| d\omega}{\int_0^\infty \omega |u_k(\omega)| d\omega} \tag{4}$$

$$\lambda^{n+1} = \lambda^n + \rho(f(t)) - \sum_k u_k^{n+1}(t)) \tag{5}$$

The solution is complete when Eq. (6) is satisfied:

$$\sum_k \frac{\|u_k^{n+1} - u_k^n\|_2^2}{\|u_k^n\|_2^2} < \varepsilon \tag{6}$$

where ε is the hyperparameter.

The detailed computational procedure for the variational modal decomposition is given in reference [23].

3.2   Multi-head Self-attention

Google Inc. Proposed Transformer for machine translation in 2017 [24]. Its internal multi-head self-attention mechanism embodies a very strong information processing capability, which is now widely used in many fields including time series prediction. The self-attention mechanism mimics the complex cognitive functions of the human brain [25], and can adaptively capture important parts of the data. The computational process of the self-attention mechanism (single-head attention) is described as follows.

First, the input matrix $X = \{x_1, x_2, \cdots, x_l\}, x_i \in \mathbb{R}^{d_{model} \times 1}$ is mapped to three feature spaces, i.e., query (Q), key (K), and value (V), and the computation procedure is described in Eq. (7):

$$Q = W^Q X, K = W^K X, V = W^V X \tag{7}$$

where W^Q, W^K and W^V are trainable parameter matrices.

Then, a scaled dot product function is used to account for its bidirectional time dependence on the historical sequence, and the attention (single-head attention) is computed as:

$$X' = Attention(Q, K, V) = Softmax(\frac{QK^T}{\sqrt{d_k}})V \tag{8}$$

In order to further enhance the feature extraction capability of the attention mechanism, an extension of the single-head attention. The multi-head self-attention mechanism allows the model to process information from several different subspaces in parallel, calculated as follows:

$$Multi - Head(Q, K, V) = Concat(head_1, \cdots, heead_n)W^O \tag{9}$$

where $head_i = Attention(WQ_i^Q, WK_i^K, WV_i^V)$

3.3 Time Encoding

In the transformer used for natural language processing, the computational process of the multi-head attention mechanism does not take into account positional information, so positional encoding of the vectors obtained from word embeddings is required. Similarly, we need to perform temporal position coding, or time coding, when we use the multi-head attention mechanism for solar irradiance prediction. The temporal encoding computation process is as follows:

$$\begin{cases} PE(t, 2i) = \sin\left(\frac{t}{10000^{2i/d_{model}}}\right) \\ PE(t, 2i + 1) = \cos\left(\frac{t}{10000^{2i/d_{model}}}\right) \end{cases} \tag{10}$$

where t is the position of the input time, i is the dimension, d_{model} is the expanded total dimension, $PE(t, 2i)$ is the coded value of the $2i$ th element, and inputs of different times and dimensions have different time position coding. The data with time-coded information can be obtained by adding the high dimensional data with the time coding.

3.4 One-Dimensional Convolutional Neural Network

CNNs are often used to analyze visual images, which are capable of fully extracting features from the image using a very small number of parameters. And in the field of time series prediction, recent studies have shown that one-dimensional convolution outperforms the two-dimensional convolution traditionally used for images in processing time series [26]. The so-called one-dimensional convolution is to replace the traditional $k * k$ convolution kernel with $k * 1$. In this paper, the model utilizes two layers of one-dimensional convolution to dimensionally expand the data, which is then inputted to the subsequent model for prediction. Where each convolution kernel is set to calculate the convolution on the time series x. The calculation process can be summarized as follows:

$$x_j' = \sum_{i=0}^{m} \sum_{j=1}^{k} a_j x_{ci+j} \tag{10}$$

where m denotes the number of steps the convolution kernel needs to move on the sequence x; k denotes the one-dimensional convolution kernel length; a is the trainable parameter of the convolution kernel; and c is the convolution kernel's step length per move.

3.5 Bidirectional Long Short Term Memory

The LSTM consists of three gate structures, the forgetting gate, the updating gate (or called the input gate), and the output gate, as shown on the left in Fig. 2. Each LSTM unit is computed as follows:

$$f_t = \sigma(W_f \cdot [h_{t-1}, x_t] + b_f) \tag{12}$$

where f_t is a forgetting gate used to decide which information from C_{t-1} is used to compute C_t, which denotes the current cell state; σ denotes the sigmoid activation function, which controls f_t between 0 and 1; W_f and b_f are trainable parameter matrices and bias coefficients; h_{t-1} is the previous cell output; and x_t is the current cell input.

$$i_t = \sigma\left(W_i \cdot [h_{t-1}, x_t] + b_i\right)$$

$$\tilde{c}_t = tanh(W_c \cdot [h_{t-1}, x_t] + b_c) \tag{13}$$

where $\tilde{c}_t$ denotes the cell state update value; tanh is the activation function; and i_t is an input gate, again a quantity between 0 and 1, used to control which information in $\tilde{c}_t$ is applied to update C_t.

$$o_t = \sigma(W_o \cdot [h_{t-1}, x_t] + b_o)$$

$$h_t = o_t * tanh(c_t) \tag{14}$$

where, o_t is the output gate to control those information in the current cell state C_t is used for output.

BiLSTM consists of two layers of forward and backward LSTMs as shown on the right side of Fig. 2, and the following computations are performed on top of the unidirectional LSTM:

$$\overrightarrow{h_t} = H\left(W_1 x_t + W_2 \overrightarrow{h_{t-1}} + \overrightarrow{b}\right)$$

$$\overleftarrow{h_t} = H\left(W_3 x_t + W_5 \overleftarrow{h_{t-1}} + \overrightarrow{b}\right)$$

$$y_t = W_4 \overrightarrow{h_t} + W_6 \overleftarrow{h_t} \tag{15}$$

where $\overrightarrow{h_t}$, $\overleftarrow{h_t}$ and y_t are the forward propagation, backward propagation and output layer vectors, respectively.

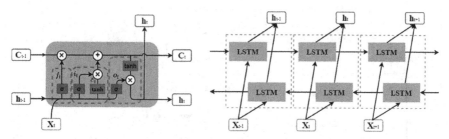

Fig. 2. LSTM cell internal structure and BiLSTM [27].

4 VMD-AC-BiLSTM

The prediction process of the proposed model is shown in Fig. 3. The prediction part of the proposed model consists of a stochastic prediction module and a periodic prediction module.

The stochastic prediction module takes a certain length of historical solar irradiance data and cloud-type data as input and outputs a prediction matrix. Cloud motion is closely related to solar irradiance, and different cloud types and weather have different patterns of influence on light. In other words, by allowing the prediction model to learn the cloud information, the prediction accuracy and prediction efficiency can be greatly improved. Therefore, the proposed model takes into account the complex effect of cloud type on the stochasticity in the solar irradiance data and inputs the cloud type data into the stochasticity prediction module as the starting state of BiLSTM.

Compared to the data in the stochastic component set, the trend component set has almost no complex stochastic fluctuations and contains only climate trends and climatic variations of relatively long duration. Therefore, the periodic prediction module is simpler and less computationally complex than the stochastic prediction module. It contains only one-dimensional CNN (1D-CNN) and BiLSTM, where 1D-CNN is used to expand the solar irradiance data dimensions.

As shown in Fig. 3, we first preprocess the raw data. The solar irradiance data is then decomposed into n sub-sequences by variational mode decomposition (VMD). The $n - 1$ sub-sequences are used as the set of stochasticity components, which are inputted into the stochasticity prediction module in the prediction section. The data in the stochasticity prediction module is first passed through a multi-head attention mechanism which adaptively highlights the important parts of the data. The data in the

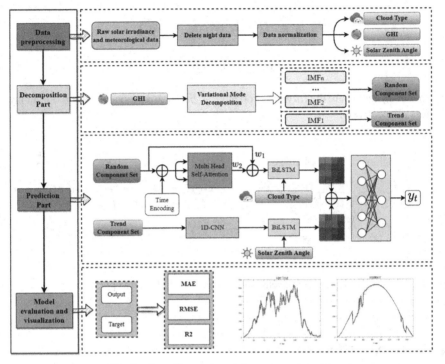

Fig. 3. Overall structure of the proposed model.

random prediction module is initially processed by the multi-head attention mechanism, which dynamically emphasizes the crucial portion of the data and assigns weights to both pre-attention and post-attention data before summing them up. The randomness prediction matrix is obtained by employing BiLSTM. The remaining 1 subsequence as the set of trend components is input to the periodicity prediction module in the prediction section. In the periodic prediction module, the data is first dimensionally expanded by 1D-CNN, and then the periodic prediction matrix is obtained by BiLSTM. The two prediction modules also consider cloud type and solar zenith angle separately. Finally, the prediction results are summarized and accumulated to get the final prediction output. Finally, the trained model is evaluated.

5 Experiment

5.1 Dataset

The dataset of this study comes from the National Solar Radiation Database (NSRDB). We collected solar irradiance and climate information for the whole year from January 1, 2021 to December 31, 2021 near San Francisco (Latitude: 37.78, Longitude: −122.41) at 5-min intervals with a collection timeframe of 365 days. Because there is no sunlight in the evening, and the sunrise and sunset times are easily available, we cropped out the part of the dataset that has no sunlight at night. The site has four distinct seasons, with relatively stable light in summer and fall, more cloudy and rainy days in winter, and drastic changes in solar irradiance, and each of the four seasons reflects a different climatic trend. Because of this, we divided the yearly data into four seasons based on the month, i.e., spring (January-March), summer (April-June), fall (July-September) and winter (October-December).

As mentioned earlier, our dataset also contains solar zenith angle and cloud type data. The solar zenith angle represents the angle between the sunlight incidence direction and the vertical direction of the zenith. The cloud type contains a total of 14 categories, which are categorized as shown in Table 1, and the details of cloud categorization can be found in the official website of PATMOS-X: https://cimss.ssec.wisc.edu/patmosx/. We preprocessed and normalized the data before the experiment, and the data of each season was divided into the training set and test set with a ratio of 4:1.

Table 1. Cloud Type Category

0	Clear	5	Mixed	10	Unknown
1	Probably Clear	6	Opaque Ice	11	Dust
2	Fog	7	Cirrus	12	Smoke
3	Water	8	Overlapping	15	N/A
4	Super-Cooled Water	9	Overshooting		

5.2 Evaluation Indicators

In this experiment, three evaluation indexes, namely mean absolute error (MAE), root mean square error (RMSE) and R-squared (R2), are selected to evaluate the performance of forecasting models. They are the most widely used evaluation indexes in time series forecasting-related research, and they can evaluate the strength or weakness of the model forecasting performance from different perspectives. Among them, both MAE and RMSE can indicate the magnitude of the error between the predicted value and the sample value, but RMSE is more sensitive to the outliers in the predicted value, while R2 indicates the degree of fitting of the predicted series. The smaller the MAE and RMSE, the better, and the larger the R2, the better, they are calculated:

$$MAE(y, \hat{y}) = \frac{1}{n} \sum_{i=1}^{n} |y_i - \hat{y}_i| \tag{11}$$

$$RMSE(y, \hat{y}) = \sqrt{\frac{1}{n} \sum_{i=1}^{n} (y_i - \hat{y}_i)^2} \tag{12}$$

$$R2(y, \hat{y}) = 1 - \frac{\sum_{i=1}^{n} (y_i - \hat{y}_i)^2}{\sum_{i=1}^{n} (\overline{y}_i - \hat{y}_i)^2} \tag{13}$$

where y is the model predicted value, $\hat{y}$ is the true value of the sample, and n is the total number of samples.

5.3 Experimental Analysis

In order to validate the predictive effectiveness and robustness of the proposed model, we have conducted a number of comparison experiments. The most prevalent deep learning models, Gated Recurrent Units (GRU, M1) and LSTM (M2), were selected for comparative experiments. The proposed model's prediction component is demonstrated to be superior by introducing a VMD-LSTM (M3) model for comparison, which utilizes the same VMD technique as the proposed model for decomposition and employs parallel LSTM layers for prediction. The proposed model's cloud type input is eliminated to verify its validity, while keeping the other components unchanged. Consequently, we obtain the VMD-AC-BiLSTM-(M4) model that does not consider the cloud type.

We trained all the models 10 times respectively under the best hyperparameter configuration. In order to avoid the influence of abnormal training to some extent, the 10 training results of all the models were removed from one prediction result with the highest error and one prediction result with the lowest error by using the MAE as the benchmark, and the prediction errors of the remaining 8 trainings of each model were averaged out. The results of the error settlements are shown in the Table 2 and the Fig. 4.

Table 2. MAE, RMSE, R^2 for solar irradiance comparison tests

		M1	M2	M3	M4	Proposed
Spring	MAE	18.2352	20.1610	19.8618	16.1779	**15.3638**
	RMSE	30.3925	31.4232	25.7942	21.8489	**20.9309**
	R^2	0.9891	0.9883	0.9919	0.9943	**0.9948**
Summer	MAE	12.5112	12.1463	13.9172	10.3892	**9.1644**
	RMSE	33.6017	33.3164	23.7767	20.8178	**18.5385**
	R^2	0.9897	0.9898	0.9947	0.9960	**0.9968**
Autumn	MAE	13.6418	13.5267	11.6807	**11.3139**	11.3923
	RMSE	41.1203	40.9904	**27.0471**	27.3226	27.6281
	R^2	0.9756	0.9757	**0.9894**	0.9892	0.9890
Winter	MAE	29.8366	30.1193	21.6311	19.1351	**18.3086**
	RMSE	51.2908	51.1257	34.5218	31.3044	**30.1878**
	R^2	0.8909	0.8916	0.9492	0.9593	**0.9621**

As can be seen from the Table 2 and Fig. 4, the proposed model achieves the best prediction results in both spring summer and winter, followed by the M4 model without considering cloud type. And the advantage of the proposed model is more obvious especially in winter. The reason for this is that the winter dataset exhibits the highest level of random volatility in solar irradiance, with a majority of this volatility concentrated within the forecast set portion of the dataset. This strong random fluctuation leads to generally poor predictions from all models. In the fall, although the proposed model predictions are not the best, they are very close to the best model prediction error and degree of fit.

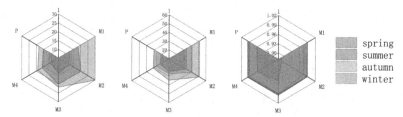

Fig. 4. Radar plot of model prediction error comparison.

This is because the solar irradiance in the fall is relatively more stable, and there are almost no very sharp and long-duration random fluctuations, so relatively simple models can achieve better prediction results. For the M3 model, it is not difficult to find that its prediction errors are mostly between those of the single deep learning model (M1, M2) and the proposed model. Because it uses VMD, this can help the subsequent prediction part to extract the effective features in the data. Its prediction accuracy improves dramatically in the winter season, when the stochastic volatility is much stronger, compared to the model without decomposition method. However, it ignores the correlation between the different subsequences, leading to no significant improvement over M1 and M2 in the spring and summer months when stochastic volatility is moderate. For the M4 model, compared with the proposed model, it only eliminates the input of cloud type, and it can be found that most of its forecast evaluation indexes are slightly worse than that of the proposed model.

We selected one day's data in each of the four seasonal test sets and visualized the real solar irradiance curves with the predicted value curves of several prediction models, and the visualization results are shown in Fig. 5. It is easy to see from the figure that each model has better prediction results in the more stable part of the data, such as around 130 points in summer and 120 points in fall. For summer and fall, M1 and M2 have more obvious time delay phenomena at the data fluctuations. This is due to the relatively simple and weak modeling ability of the M1 and M2 models, whereas there are fewer random fluctuations in the summer and fall datasets, which leads to their stronger dependence on the proximity points. Whereas for the M3, M4, and the proposed models, although the time delay phenomena are not obvious, at the data fluctuations, especially at the part of the sudden changes in the solar irradiance, the They all show different degrees of prediction bias. This could be due to the fact that there are fewer fluctuating situations in the training set, resulting in inexperience of the models in coping with the fluctuations. Whereas the proposed model takes into account the cloud type information, so the bias phenomenon is weaker than that of M3 and M4, and reflects a stronger robustness. For spring and winter, when stochastic fluctuations are stronger, especially around 85 o'clock in winter, all models show large prediction errors, with M1 and M2 showing the worst prediction. The implication is that a solitary model is insufficient to handle such unpredictable fluctuations. Not only this, but also M3 shows prediction anomalies near 90 o'clock in winter. This suggests that prediction methods that discard correlations between subsequences are more unstable. Whereas the M4 and proposed models, although they also have prediction bias, have improved significantly compared to the other models.

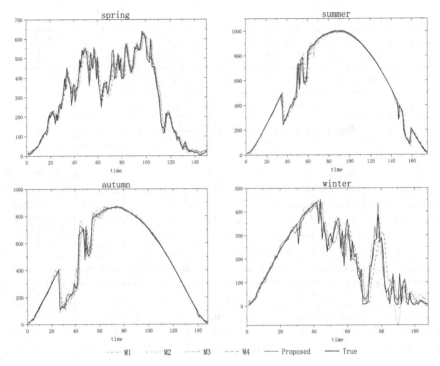

Fig. 5. Visualization of the results of the four seasons forecast.

As mentioned earlier, we trained each model 10 times at the optimal hyperparameter settings, we analyze the stability of the models. The MAE violin plots for several models over the four seasons are shown in Fig. 6. The outermost curve of each color-coded shape in the figure represents the probability distribution of the error, with a narrower longitudinal span indicating a higher level of prediction stability.

As can be seen from the figure, in the spring dataset, M1 has the most stable prediction effect, M3 shows an approximately uniform distribution over a wide range, while M4 has more obvious anomalous training. The proposed model has a generally lower MAE value, although it has a wider distribution than M1. For the summer season, all model prediction errors remain within a low range, with the proposed model having the lowest error, followed by M4, which uses the same structure as the proposed model, and M3 having a very clear anomalous training. Each model in the summer dataset exhibits similar distributional characteristics to the fall dataset, with only M4 having a more pronounced anomalous training. As for winter, the proposed model has the most obvious prediction advantage. The prediction errors of M4 with the proposed model are significantly lower than those of M1 and M2. M3 can achieve similar prediction results with the proposed model, but some of the prediction errors are at a higher level.

In summary, the proposed model shows the best prediction effect in all four seasons and demonstrates high prediction stability. M3 shows very obvious anomalous prediction in summer, and the prediction errors are widely distributed in spring and winter.

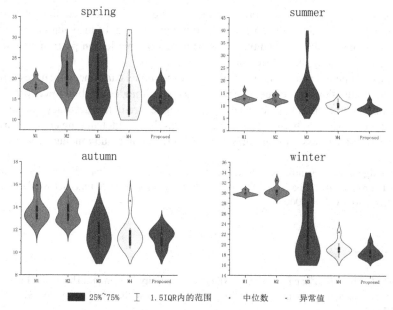

Fig. 6. Model Prediction Error Comparison Violin Plot.

The instability of the predictive effects may be attributed to the disregard for the interdependencies among subsequences in different frequency ranges, thereby compromising its effectiveness. M4 only eliminates the input of cloud type information compared with the proposed model, so it presents a slightly worse prediction effect than the proposed model in the four seasons. But M4 shows very obvious anomalous prediction in spring and fall. Therefore, we believe that the inclusion of cloud type information can slightly improve the prediction accuracy of the model, but at the same time, it can significantly improve the stability of the model prediction.

6 Conclusion

In this paper, we use historical solar irradiance data and cloud type and solar zenith angle data to make ultra-short-term accurate predictions of future solar irradiance. Using the NSRDB public solar irradiance dataset, we propose a decomposition-integrated deep learning model named VMD-AC-BiLSTM through an in-depth study of the data characteristics. The model firstly decomposes the historical data series into several subsequences by VMD, and then divides these subsequences into a stochastic component set and a trending component set. The two sets are respectively inputted into the stochastic prediction module and the periodic prediction module. The stochastic prediction module has higher complexity and stronger data modeling capability. It contains a multi-head self-attention mechanism and a bidirectional long- and short-term memory network. At the same time it takes into account the complex effect of cloud type on the stochasticity of solar irradiance. The periodic prediction module is relatively simple and has lower computational complexity. It contains a one-dimensional convolutional neural network and a bidirectional long- and short-term memory network. It predicts the solar irradiance through the solar zenith angle and the trend component of the solar periodicity in irradiance. Meanwhile, we quantitatively evaluated the prediction performance of the proposed model using three different evaluation criteria.

The experimental results show that the proposed model can cope with more complex stochastic fluctuations more accurately. The proposed model exhibits lower prediction error, more accurate fitting of the real data curve, higher prediction accuracy, and stronger robustness compared to both the deep learning baseline model and the control variable correction model across all four datasets of spring, summer, autumn, and winter.

Acknowledgement of Fundings. The research was funded by the foundation project: National Key R&D Program of China (No. 2021YFC3340400) and Zhejiang Natural Science Foundation Committee (No. LQ20F050009).

References

1. Guo, Z., Zhou, K., Zhang, C., et al.: Residential electricity consumption behavior: Influencing factors, related theories and intervention strategies. Renew. Sustain. Energy Rev. **81**, 399–412 (2018)
2. Bah, M.M., Azam, M.: Investigating the relationship between electricity consumption and economic growth: evidence from South Africa. Renew. Sustain. Energy Rev. **80**, 531–537 (2017)
3. Chae, Y.J., Lee, J.I.: Thermodynamic analysis of compressed and liquid carbon dioxide energy storage system integrated with steam cycle for flexible operation of thermal power plant. Energy Convers. Manage. **256**, 115374 (2022)
4. Carley, S., Baldwin, E., MacLean, L.M., et al.: Global expansion of renewable energy generation: an analysis of policy instruments. Environ. Resource Econ. **68**, 397–440 (2017)
5. Scolari, E., Reyes-Chamorro, L., Sossan, F., et al.: A comprehensive assessment of the short-term uncertainty of grid-connected PV systems. IEEE Trans. Sustainable Energy **9**(3), 1458–1467 (2018)
6. Wang, W., Chen, H., Lou, B., et al: Data-driven intelligent maintenance planning of smart meter reparations for large-scale smart electric power grid. In: 2018 IEEE SmartWorld, Ubiquitous Intelligence & Computing, Advanced & Trusted Computing, Scalable Computing & Communications, Cloud & Big Data Computing, Internet of People and Smart City Innovation, pp. 1929–1935. IEEE (2018)
7. Huang, X., Li, Q., Tai, Y., et al.: Time series forecasting for hourly photovoltaic power using conditional generative adversarial network and Bi-LSTM. Energy **246**, 123403 (2022)
8. Yona, A., Senjyu, T., Funabashi, T., et al.: Optimizing re-planning operation for smart house applying solar radiation forecasting. Appl. Sci. **4**(3), 366–379 (2014)
9. Pi, M., Jin, N., Ma, X., et al.: Short-term solar irradiation prediction model based on WCNN_ALSTM. In: 2021 IEEE Intl Conf on Dependable, Autonomic and Secure Computing, Intl Conf on Pervasive Intelligence and Computing, Intl Conf on Cloud and Big Data Computing, Intl Conf on Cyber Science and Technology Congress, pp. 405–412. IEEE (2021)
10. Zhang, L., Wang, J., Niu, X., et al.: Ensemble wind speed forecasting with multi-objective Archimedes optimization algorithm and sub-model selection. Appl. Energy **301**, 117449 (2021)
11. Jin, N., Yang, F., Mo, Y., et al.: Highly accurate energy consumption forecasting model based on parallel LSTM neural networks. Adv. Eng. Inform. **51**, 101442 (2022)

12. Li, Y., Zhu, Z., Kong, D., et al.: EA-LSTM: evolutionary attention-based LSTM for time series prediction. Knowl.-Based Syst.Based Syst. **181**, 104785 (2019)
13. Li, Q., Zhang, D., Yan, K.: A solar irradiance forecasting framework based on the CEE-WGAN-LSTM model. Sensors **23**(5), 2799 (2023)
14. Singla, P., Duhan, M., Saroha, S.: An ensemble method to forecast 24-h ahead solar irradiance using wavelet decomposition and BiLSTM deep learning network. Earth Sci. Inf. **15**(1), 291–306 (2022)
15. Cao, J., Li, Z., Li, J.: Financial time series forecasting model based on CEEMDAN and LSTM. Physica A **519**, 127–139 (2019)
16. Benidis, K., Rangapuram, S.S., Flunkert, V., et al.: Deep learning for time series forecasting: tutorial and literature survey. ACM Comput. Surv.Comput. Surv. **55**(6), 1–36 (2022)
17. Bandara, K., Bergmeir, C., Smyl, S.: Forecasting across time series databases using recurrent neural networks on groups of similar series: a clustering approach. Expert Syst. Appl. **140**, 112896 (2020)
18. Qing, X., Niu, Y.: Hourly day-ahead solar irradiance prediction using weather forecasts by LSTM. Energy **148**, 461–468 (2018)
19. Cascone, L., Sadiq, S., Ullah, S., et al.: Predicting household electric power consumption using multi-step time series with convolutional LSTM. Big Data Research **31**, 100360 (2023)
20. Guo, J., Wang, W., Tang, Y., et al.: A CNN-Bi_LSTM parallel network approach for train travel time prediction. Knowl.-Based Syst. **256**, 109796 (2022)
21. Zeng, Y., Chen, J., Jin, N., et al.: Air quality forecasting with hybrid LSTM and extended stationary wavelet transform. Build. Environ. **213**, 108822 (2022)
22. Pi, M., Jin, N., Chen, D., et al.: Short-term solar irradiance prediction based on multichannel LSTM neural networks using edge-based IoT system. Wirel. Commun. Mob. Comput. **2022**, 1–11 (2022)
23. Dragomiretskiy, K., Zosso, D.: Variational mode decomposition. IEEE Trans. Signal Process. **62**(3), 531–544 (2013)
24. Vaswani, A., Shazeer, N., Parmar, N., et al.: Attention is all you need. Advances in neural information processing systems, 30 (2017)
25. Li, R., Zeng, D., Li, T., et al.: Real-time prediction of SO2 emission concentration under wide range of variable loads by convolution-LSTM VE-transformer. Energy **269**, 126781 (2023)
26. Markova, M.: Convolutional neural networks for forex time series forecasting. In: AIP Conference Proceedings. AIP Publishing 2459(1) (2022)
27. Wang, H., Zhang, Y., Liang, J., et al.: DAFA-BiLSTM: deep autoregression feature augmented bidirectional LSTM network for time series prediction. Neural Netw. **157**, 240–256 (2023)

Anomaly Detection of Industrial Data Based on Multivariate Multi Scale Analysis

Dan Lu, Siao Li, Yingnan Zhao[✉], and Qilong Han

College of Computer Science and Technology,
Harbin Engineering University, Harbin 150009, China
{luan,zhaoyingnan,hanqilong}@hrbeu.edu.cn

Abstract. Anomaly detection stands as a crucial facet within the domain of data quality assurance. Notably, significant strides have been made within the realm of existing anomaly detection algorithms, encompassing notable techniques such as Long Short-Term Memory (LSTM), Gated Recurrent Unit (GRU), and anomaly detection models founded upon Generative Adversarial Networks (GANs). However, a notable gap lies in the inadequate consideration of interdependencies and correlations inherent in multidimensional time-series data. This becomes particularly pronounced within the context of industrial evolution, where industrial data burgeons in complexity. To address this lacuna, a novel hybrid model has been introduced, synergizing the capabilities of GRU with structural learning methodologies and graph neural networks. The model capitalizes on graph structural learning to unearth dependencies linking data points across distinct spatial dimensions. Concurrently, GRU extracts temporal correlations embedded within data along a single dimension. Through the incorporation of graph attention networks, the model employs a dual-faceted correlation perspective for data prediction. Discrepancies between predicted values and ground truth are utilized to gauge errors. The amalgamation of predictive and scoring mechanisms enhances the model's versatility. Empirical validation on two authentic sensor datasets unequivocally demonstrates the superior efficacy of this approach in anomaly detection compared to alternative methodologies. A notable augmentation is observed particularly in the recall rate, underscoring the method's potency in identifying anomalies.

Keywords: Graph neural network · Multi-source correlation · GRU · Abnormally detection

1 Introduction

Anomaly detection plays a pivotal role in industrial data, with far-reaching implications for ensuring production safety, enhancing efficiency, and reducing costs. Data in the industrial domain spans various aspects from production equipment

H. Jin et al. (Eds.): GPC 2023, LNCS 14503, pp. 88–100, 2024.
https://doi.org/10.1007/978-981-99-9893-7_7

and supply chains to product quality. Anomalies within this data often signify potential issues or machinery malfunctions. The significance of anomaly detection manifests in several key aspects. Primarily, anomaly detection facilitates the early identification of anomalies within equipment or production processes, such as machinery failures or disruptions in the supply chain. Through real-time monitoring and data analysis, anomaly detection systems can capture deviations from normal patterns, enabling timely interventions before issues propagate and preempting potential production interruptions and quality concerns. Furthermore, anomaly detection significantly contributes to the enhancement of production efficiency. By monitoring data along production lines and identifying anomalies, engineers can pinpoint underlying causes of potential issues. Timely adjustments of parameters and process optimization can effectively minimize resource wastage and energy consumption during production, leading to heightened efficiency. In essence, the role of anomaly detection in industrial data is pivotal, safeguarding operations, driving efficiency gains, and ultimately supporting the overarching goal of sustainable development within the industrial sector.

There are three problems for anomaly detection: 1. It needs to combine domain knowledge [1]; Data anomaly detection needs to emphasize the deep integration of causality, domain knowledge and data analysis process. In the face of complex scene rules contained in data, insufficient scene investigation and preparation often lead to an increase in the probability of failure of the designed anomaly detection method in application. 2. Lack of appropriate datasets [1–4]; Big data is composed of traditional relational data and temporal data, but there are many problems in data such as insufficient annotation and imbalance between classes. However, the artificial intelligence method of anomaly detection is often strict in the preset of model training data, and the existence of biased heterogeneous data makes the model training very difficult. 3. Large scale and complex model [5–8]; There are some hidden correlations between data, and the patterns of big data and ordinary data are not interconnected.

The existing anomaly detection models are mainly divided into traditional algorithms and deep learning algorithms.

Traditional algorithms, such as cluster-based, proximity based, binary classification algorithm, etc., have also been widely used in some scenarios due to the improvement of existing researchers. For example, in the partition clustering algorithm, the density function is fused to select the optimal cluster center to avoid the dilemma of local minimization; For the proximity based algorithm, the reverse K-nearest neighbor is used to redetermine the K-nearest neighbor of the abnormal cluster, which also avoids local optimization. These traditional algorithms cannot play their advantages in the context of big data. Large-scale data will lead to great complexity of the algorithm. Therefore, the deep learning algorithm is developed in the big data scenario.

At present, the deep learning algorithms for solving the anomaly detection of sequential data, such as LSTM, self-coder and other models, have developed very rapidly. Because of the characteristics of data timing, LSTM (short and long term memory network) has been widely used and produced many variations. For example, the GAN model based on LSTM applies LSTM to the generator and discriminator that generate the confrontation network, and adds its time correlation to the data [9]; Stacked LSTM plus full connection layer for data prediction; Using the same encoding and decoding architecture as the self-coder, its neurons are replaced with multiple LSTM units. There are many improved applications that utilize the generation of confrontation networks, Lee [10] proposed a framework to compare different models based on generating confrontation networks, concluded that anomaly detection based on GAN has a good prediction performance, and found that adjusting the number of back-propagation of back-mapping technology can improve the prediction performance. Bashar et al. [11] used the Generative adversarial network to complete the task of anomaly detection in the extreme case that only a few data are available. Xu L [12] and others put forward a generation countermine network model based on transformer for the problem that GAN cannot process high-dimensional data. The transformer module is used to replace the generator and discriminator of GAN, and the loss function of the generator and discriminator is used as the abnormal score, which ultimately has a good efficiency. Amarbayasgalan T [3]. proposed a new unsupervised anomaly detection method for time series data based on deep learning. The model consists of two modules: time series reconstructor and anomaly detector. The time series reconstructor module uses the autoregressive model to find the optimal window width, and prepares the subsequence according to the width for further analysis. Then, it uses the depth automatic encoder model to learn the data distribution, and then uses the data distribution to reconstruct the near-normal time series. Wu [13] et al. utilized stacked GRUs to process temporal data with seasonal features and developed an anomaly detection method using matrix operations.

For solving the problem of correlation, there are many models using graph neural network to extract the correlation, Deng A [4] et al. proposed a method combining structural learning method and graph neural network, and used attention weight to provide the interpretation of detected anomalies, but the accuracy of anomaly detection was not significantly improved compared with the generation of confrontation networks. Guan, S [5] and others proposed to extract spatial correlation using graph attention network and temporal correlation using time convolution network, and finally reconstruct and predict the data. The result shows that the effect of anomaly detection has been greatly improved, but the complexity is high and the efficiency is low.

Therefore, this paper proposes a model that combines GRU, graph structure learning, and graph neural networks. This model can obtain multiple correlations while obtaining temporal correlations, and has better results for data with complex hidden relationships. The contributions of this article are summarized as follows:

– Firstly, use graph structure learning to process data, process each node into multiple feature vectors to represent the adjacency relationship among them, and then use GRU to extract temporal features from nodes with new features and splice them with the original features.

– Add a residual module to achieve depth feature extraction, while preventing gradient disappearance, gradient explosion, over fitting, and network degradation, accelerating model convergence

2 Related Work

In this chapter, we introduce the relevant content of anomaly detection for multivariate temporal data, introduce the concepts and common methods of temporal correlation and multivariate correlation, and focus on the following figure of neural network processing for multivariate data.

2.1 Multivariate Time Series Anomaly Detection

Anomaly detection is the identification of items, events or observations that do not match the expected pattern or other items in the data set. Anomalies are also referred to as outliers, novelties, noise, biases, and exceptions. Classical methods include traditional classical algorithms, heuristic algorithms, deep learning methods, etc. The current considerations of these algorithms are not comprehensive for multidimensional time-series data, including two dimensions of relevance, the first is on the temporal dimension, for which deep learning has made good progress, and the second is on the spatial dimension, for example, Chen et al. [14] conducted spatial and temporal anomaly detection in the time slice anomaly detection module, graph neural networks, graph attention networks The graph neural network, the graph attention network, has also performed very well in this dimension. Therefore, our goal is to use graph attention networks and recurrent neural networks together to complete the processing of multidimensional temporal data in two dimensions.

2.2 Time Correlations

For time-series data, temporal correlation is the most important factor, and deep learning models are widely used in large-scale time-series data processing. Nowadays, the mainstream models such as, GAN, VAE, LSTM, GRU, Transformer, and other variants and synthesis of these models are more widely used and mature, and Guo Y et al. [15] proposed an anomaly detection for GRU-based Gaussian hybrid VAE system, where GRU units are used to discover correlations between time series. A Gaussian mixed prior in the potential space is then used to characterize multimodal data. Tang C et al. [16] proposed GRN, an interpretable multivariate time series anomaly detection method based on neural graph networks and gated recurrent units GRU, which retains the original advantages of processing sequences and capturing time series correlations, and addresses the problems of gradient disappearance and gradient explosion. However, these models perform well for time series only, and still take the form of parallel processing for multidimensional time series.

2.3 Multiple Correlations

Multiple correlation means for having correlation and dependency between different dimensions in multiple dimensional data, for example, in a water tank, the data collected by the upper pressure sensor and the lower sensor is a pair of positive correlation, for dealing with multiple correlation, there are many methods, for example, using matrix to process data, self-attention mechanism, graph neural network, etc., among which graph neural network is more effective for modeling complex patterns More effective, graph neural network assumes that there are connections between nodes, and represents the characteristics of nodes by aggregating the neighbors between nodes, based on graph neural network, graph attention network uses attention mechanism to calculate the weights between neighboring nodes. Where the input of the graph neural network should be a graph structure, but the data we use is an $n \times m$ matrix in which there are no connections, so the graph neural network is preceded by using the graph structure learning method GRU to compose a list of temporal and spatial features for the graph neural network training and prediction.

3 Method

In this chapter, the problems to be solved and the model architecture are introduced, including the entire process of feature extraction, prediction, and scoring of data.

3.1 Problem Definition

The time series comprises observations at regularly spaced time intervals, and our research objective is a multivariate time series denoted as $x \in R^{N \times K}$, where N represents the length of the time series and K represents the dimension of each sample at a given moment. The training set, $x_T \in R^{M \times K}$, is used where M (M < N) represents the length of the training set, and the remaining data serves as the testing set. The training set consists solely of normal samples, while the testing set includes both normal and abnormal samples. The model takes input in the form of a sliding window of data, denoted as $x_L \in R^{K \times L}$, where L represents the length of the sliding window. We utilize the graph deviation score as a criterion for identifying anomalies. If the score surpasses the threshold, the sample is labeled as an anomaly; otherwise, it is classified as normal (Fig. 1).

3.2 Model Architecture

The model is divided into three parts: feature extraction, prediction and scoring.

- Feature extraction:Use graph structure learning to determine the correlation between nodes and extract node spatial features; use GRU to process data and obtain local features and temporal information features of time series

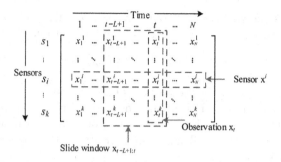

Fig. 1. Data formulation of the multivariate time series.

- Prediction: Using Graph Attention Networks for Predicting Graph Data with Temporal and Correlation Features
- Scoring: Using the graph deviation scores as the final anomaly scores, the errors for each dimension are calculated for the final merge, and to prevent the deviations caused by any one sensor from being overly dominant relative to the others, we perform a robust normalization of the error values for each sensor (Fig. 2)

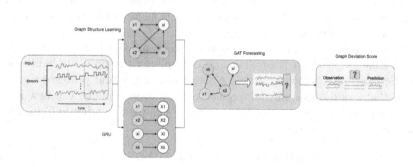

Fig. 2. Overall framework of anomaly detection model for multivariate time series

Feature Extraction. Feature extraction in the model is divided into extracting time features and correlation features, where GRU is utilized to extract time features, and graph structure learning is used to construct correlation graphs representing relationships among different dimensions.

Graph Structure Learning

The main objective of the first part of feature extraction is to learn the relationship between sensors in the form of graph structures. Initially, the raw data is aggregated and then modeled using the normalized dot product of the behavioral features of each dimension with respect to other dimensions. An adjacency matrix A is constructed to store strongly correlated adjacent dimensions, where A^{ij} denotes the existence of a connection from node i to node j.

$$A^{ij} = 1(j \in edgeindex) \tag{1}$$

We embed the normalized dot product (e^{ij}) between dimension i and the vectors of other dimensions into node i. m such normalized dot products are selected, where m refers to the top m dimensions with strong correlation, and the value of m can be adjusted to meet different requirements for the sparsity of the graph specified by the user.

$$e^{ij} = \frac{V^{i^{T}V^{j}}}{\|V^i\| \|V^j\|} \tag{2}$$

Gated Recurrent Unit
The main objective of the second part of feature extraction is to extract the temporal features of the data using GRU. In GRU, the reset gate r_t and update gate z_t are used to control the interaction between the input x_t and the previous hidden state h_{t-1}. Therefore, the time features can be obtained by merging the input of the current time step and the hidden state of the previous time step, and then embedding the resulting hidden vector h_t into node i at time t.

Reset gate: $r_t = \sigma(W_r[x_t, h_{t-1}])$. Update gate: $z_t = \sigma(W_z[x_t, h_{t-1}])$. By using reset gate and update gate, the model can choose whether to retain the previous information or update it. Here, σ represents the sigmoid function, and W_r and W_z are weight matrices.

Based on the reset gate, we can compute the candidate hidden state $\tilde{h}$ at the current time step, which is obtained by multiplying the previous hidden state with the reset gate, allowing the time step to intervene and preserve the information that could affect the subsequent computation. Finally, the new hidden state h_t is calculated as the weighted average of the new candidate hidden state $\tilde{h}$ and the past hidden state.

$$\tilde{h} = tanh(x_t W_{hx} + r_t \odot h_{t-1} W_{hh} + b_h) \tag{3}$$

$$h_t = (1 - z_t) \odot h_{t-1} + z_t \odot \tilde{h_t} \tag{4}$$

Prediction. To identify whether the data is anomalous, we employ a predictive method to forecast future time points and determine if the actual values deviate from the predicted values. The preprocessed data is transformed into a graph structure with time features, where each node possesses a set of features and interacts with other nodes. The interactions should be meaningful, with more attention given to important node interactions. Therefore, a graph attention network is utilized to calculate the importance of each node by an algorithm and generate representations for each node based on its importance.

Graph Attention Network

The input of a GAT layer is a set of node features $h = \left\{ \vec{h_1}, \vec{h_2}, \ldots, \vec{h_k}, \vec{h_i} \in R^F \right\}$, where k represents the number of nodes and F represents the number of features per node, and the output is a set of new node features $h' = \left\{ \vec{h_1'}, \vec{h_2'}, \ldots, \vec{h_k'}, \vec{h_i'} \in R^{F'} \right\}$. To obtain sufficient expressiveness and transform the input features into higher-level representations, at least one learnable linear transformation is required. Therefore, as a first step, a shared linear transformation, parameterized by a weight matrix, $W \in R^{F' \times F}$, is applied to each node. Then, a self-attention mechanism, a shared attention mechanism, is applied to the nodes to compute the attention coefficients.

$$e_{ij} = a(W\vec{h_i}, W\vec{h_j}) \tag{5}$$

e_{ij} represents the importance of the characteristics of node j to node i. In GAT, each node is allowed to participate in the activity of other nodes by injecting the graph structure into the mechanism using a masked attention mechanism. The coefficient e_{ij} is computed for node i's neighbor j, in the graph, this corresponds to the first-order neighborhood of node i (including i). To make the coefficients easily comparable across different nodes, we normalize all j options using the softmax function and use LeakyReLU as the non-linear activation function to compute attention coefficients.

$$a_{ij} = softmax_j(e_{ij}) = \frac{exp(e_{ij})}{\sum_{k(A^{ik}>0)} exp(e_{ik})} \tag{6}$$

$$a_{ij} = \frac{exp(LeakyReLU(\vec{a}^T[W\vec{h_i}||W\vec{h_j}]))}{\sum_{k(A^{ik}>0)} exp(LeakyReLU(\vec{a}^T[W\vec{h_i}||W\vec{h_j}]))} \tag{7}$$

With the normalization of attention coefficients, the final output feature $\vec{h_1'}$ for each node can be computed. Subsequently, the results from all nodes are utilized as input to a stacked fully connected layer with an output dimension of N, for the purpose of predicting the vector of sensor values at time step t, denoted as x_{pre}. The mean square error between the predicted output and observed data is employed as the loss function.

$$L_{MSE} = \frac{1}{T_{train} - w} \sum_{T_{train}}^{t=w+1} \left\| x_{pre}^{(t)} - x^{(t)} \right\|_2^2 \tag{8}$$

Scoring. Our model computes individual anomaly scores for each dimension and combines them to form a single anomaly score for each time step, enabling the user to identify which sensors are anomalous. The anomaly score compares the expected behavior at time t with the observed behavior, calculating the error fall at time t and dimension i. Due to the differing magnitudes of data fluctuations across each dimension, in order to mitigate the influence of certain dimensions on the overall result, we normalize the error values for each dimension:

$$fal_i(t) = \left| x_i^t - x_{pre_i}^t \right| \tag{9}$$

$$ad_i(t) = \frac{fal_i(t) - \tilde{\mu}_i}{\tilde{\sigma}_i} \tag{10}$$

where $\tilde{\mu}_i$ and $\tilde{\sigma}_i$ are the median and interquartile range of the $fal_i(t)$ values at the time step. We use the median and instead of the mean and interquartile range standard deviation, as they are more robust to outliers.

Subsequently, we simply select the maximum value as the overall anomaly at time t, as anomalies are rare events that may only affect a small subset of dimensions. Finally, we use a simple moving average (SMA) to generate a smoothed score $As(t)$, and set a fixed threshold for anomalies to be detected when $As(t)$ exceeds the threshold value.

$$AD(t) = \max_i ad_i(t) \tag{11}$$

3.3   Residual Network

We incorporated a residual network into the graph attention network to address the problem of network degradation in multi-layer neural networks. Forward information transmission filters out irrelevant information, while backward information transmission represents an error. In fact, as the network depth increases, the training error also increases. Excessive network layers lead to identity mapping in subsequent layers, causing resource waste.

$$y_l = h(x_i) + F(x_i, W_l) \tag{12}$$

4   Experiment

In this chapter, the relevant contents of the experiment are introduced, including the data set experiment results and result analysis.

4.1   Datasets

We utilized two sensor datasets related to water treatment: SWaT and WADI. The data size and composition of the datasets are presented in Table 1. Among them, Swat contains 92501 pieces of data, and WADI contains 136070 pieces of data.

The Secure Water Treatment (SWaT) dataset comes from a water treatment test-bed coordinated by Singapore's Public Utility Board (Mathur and Tippenhauer 2016). It represents a small-scale version of a realistic modern Cyber-Physical system, integrating digital and physical elements to control and monitor system behaviors. As an extension of SWaT, Water Distribution (WADI) is a distribution system comprising a larger number of water distribution pipelines (Ahmed, Palleti, and Mathur 2017). Thus WADI forms a more complete and realistic water treatment, storage and distribution network. The datasets contain two weeks of data from normal operations, which are used as training data for the respective models. A number of controlled, physical attacks are conducted at different intervals in the following days, which correspond to the anomalies in the test set.

Table 1. Statistics of the two datasets used in experiments.

Datasets	Features	Anomalies
SWaT	51	11.97%
WADI	127	5.99%

4.2   Baselines

We have compared the performance of this method with several popular anomaly detection methods and the original method, including:

AE: Automatic encoder, wherein the encoder is used for dimensionality reduction of data, and the decoder is used for dimensionality enhancement to complete data reconstruction, using the reconstruction error as an abnormality score.

DAGMM: [17] The automatic encoding Gaussian model combines a depth automatic encoder and a Gaussian mixture model, using the reconstruction error of the self encoder and the likelihood function of the Gaussian model to determine anomalies.

LSTM-VAE: [18] The variational self encoded hidden variable is used as the input to LSTM to predict the next hidden variable, and the prediction error is finally taken as an outlier.

MAD-GAN: [19] Using the LSTM layer as the generator and discriminator of the GAN model, anomaly detection is performed through prediction and reconstruction methods.

GDN: [4]Using Graph Structure Learning and Graph Attention Network for Correlation Learning to Predict and Judge Anomalies.

4.3   Experimental Setup

We implement our method and its variants in PyTorch version 1.5.1 with CUDA 10.2 and PyTorch Geometric Library version 1.5.0, and train them on a server with Intel(R) Core(TM) i7-8750H CPU @ 2.20 GHz and 4 NVIDIA RTX 1060 graphics cards. The models are trained using the Adam optimizer with learning rate 1×10^{-3} and $(\beta 1, \beta 2) = (0.9, 0.99)$. We train models for up to 30 epochs. We use embedding vectors with length of 64, k with 15 and hidden layers of 64 neurons for the datasets, corresponding to their difference in input dimensionality. We set the sliding window size w as 5 for both datasets.

4.4   Evaluation Metrics

We employ precision (Prec), recall (Rec), and F1-Score (F1) to evaluate our approach and baseline models, where $F1 = \frac{2 \times Prec \times Rec}{Prec + Rec}$, $Prec = \frac{TP}{TP+FP}$, and $Rec = \frac{TP}{TP+FN}$. Here, TP, TN, FP, and FN represent the numbers of true positives, true negatives, false positives, and false negatives, respectively. To detect anomalies, we set the threshold using the maximum anomaly score on the validation dataset. During testing, any time step exceeding the threshold will be classified as an anomaly.

4.5   Results

Table 2. Anomaly detection accuracy in terms of precision (%), recall (%), and F1-score, on two datasets with ground-truth labelled anomalies.

Method	SWAT			WADI		
	Prec	Rec	F1	Prec	Rec	F1
AE	72.63	52.63	0.61	34.35	34.35	0.34
DAGMM	27.46	69.52	0.39	54.44	26.99	0.36
LSTM-VAE	96.24	59.91	0.74	87.79	14.45	0.25
MAD-GAN	98.97	63.74	0.77	41.44	33.92	0.37
GDN	99.75	58.12	0.72	78.65	34.03	0.47
GRU-GDN	**95.84**	**72.60**	**0.82**	**87.82**	**35.03**	**0.49**

In Table 2, we show the anomaly detection accuracy in terms of precision, recall and F1-score, of our GRU-GDN method and the baselines, on the SWaT and WADI datasets. The results show that GRU-GDN outperforms the baselines in both datasets, with high precision in both datasets of 0.95 on SWaT and 0.87 on WADI. The recall rate on SWaT has increased significantly, by 14%, and the accuracy rate on WADI has increased significantly, by 9%.

In the context of industrial data, the emphasis lies in effectively detecting genuine anomalies, making the recall rate a crucial metric of interest. Specifically, concerning the SWaT dataset, the GRU-GDN model exhibits superior performance in terms of recall. Additionally, when dealing with the more complex and voluminous WADI dataset characterized by higher dimensions, the GRU-GDN model demonstrates a comprehensive improvement across all three evaluation metrics. These outcomes underscore the significance of the model's ability to extract temporal correlations and multidimensional spatial relationships, thereby enhancing the accuracy of anomaly detection. Furthermore, these results provide a substantiated rationale for the observed superiority of the GRU-GDN model over the baseline techniques.

4.6   Ablation Experiment

In order to study the effectiveness of the method in improving the effectiveness of anomaly detection, we separately excluded these components to observe the decline in model performance. First, we replace GRU with LSTM to extract temporal features. Secondly, we eliminate the attention mechanism and assign the same weight to different nodes. The results are summarized as follows(Table 3):

Table 3. Anomaly detection accuracy in terms of precision (%), recall (%), and F1-score.

Method	SWAT			WADI		
	Prec	Rec	F1	Prec	Rec	F1
GRU-GDN	**95.84**	**72.60**	**0.82**	**87.82**	**35.03**	**0.49**
LSTM-GDN	95.12	66.57	0.78	67.95	35.34	0.46
-attention	72.63	52.63	0.61	85.50	29.61	1.43

5   Conclusions

In this work, we use the framework of graph structure learning and graph attention networks to complete the extraction and prediction of spatial features. We propose to combine GRU to extract temporal features in parallel, and complete data anomaly detection tasks with long time series, complex features, and multiple dimensions. Experimental validation on data sets shows that the GRU-GDN method outperforms the baseline method in terms of accuracy, recall, and F-score, Future work can simplify the model and use it online, improving the practicality of the method

Acknowledgments. This work was supported by the National Key R&D Program of China under Grant No. 2020YFB1710200. The datasets are provided by iTrust, Centre for Research in Cyber Security, Singapore University of Technology and Design.

References

1. Zhao, P., Chang, X., Wang, M.: A novel multivariate time-series anomaly detection approach using an unsupervised deep neural network. IEEE Access **9**, 109025–109041 (2021)
2. Schmidl, S., Wenig, P., Papenbrock, T.: Anomaly detection in time series: a comprehensive evaluation. Proc. VLDB Endow. **15**(9), 1779–1797 (2022)

3. Amarbayasgalan, T., Pham, V.H., Theera-Umpon, N., et al.: Unsupervised anomaly detection approach for time-series in multi-domains using deep reconstruction error. Symmetry **12**(8), 1251 (2020)
4. Deng, A., Hooi, B.: Graph neural network-based anomaly detection in multivariate time series. In: Proceedings of the AAAI Conference on Artificial Intelligence, vol. 35, no. 5, pp. 4027–4035 (2021)
5. Guan, S., Zhao, B., Dong, Z., Gao, M., He, Z.: GTAD: graph and temporal neural network for multivariate time series anomaly detection. Entropy **24**(6), 759 (2022)
6. Zhou, H., Yu, K., Zhang, X., et al.: Contrastive autoencoder for anomaly detection in multivariate time series. Inf. Sci. **610**, 266–280 (2022)
7. Almardeny, Y., Boujnah, N., Cleary, F.: A novel outlier detection method for multivariate data. IEEE Trans. Knowl. Data Eng. 1 (2020). https://doi.org/10.1109/tkde.2020.3036524
8. Pasini, K., Khouadjia, M., Samé, A., et al.: Contextual anomaly detection on time series: a case study of metro ridership analysis. Neural Comput. Appl. **34**(2), 1483–1507 (2022)
9. Niu, Z., Yu, K., Wu, X.: LSTM-based VAE-GAN for time-series anomaly detection. Sensors **20**(13), 3738 (2020). https://doi.org/10.3390/s20133738
10. Lee, C.K., Cheon, Y.J., Hwang, W.Y.: Studies on the GAN-based anomaly detection methods for the time series data. IEEE Access **9**, 73201–73215 (2021)
11. Bashar, M.A., Nayak, R.: TAnoGAN: time series anomaly detection with generative adversarial networks. In: 2020 IEEE Symposium Series on Computational Intelligence (SSCI), pp. 1778–1785. IEEE (2020)
12. Xu, L., Xu, K., Qin, Y., et al.: TGAN-AD: transformer-based GAN for anomaly detection of time series data. Appl. Sci. **12**(16), 8085 (2022)
13. Wu, W., He, L., Lin, W., et al.: Developing an unsupervised real-time anomaly detection scheme for time series with multi-seasonality. IEEE Trans. Knowl. Data Eng. (2020)
14. Chen, L.J., Ho, Y.H., Hsieh, H.H., et al.: ADF: an anomaly detection framework for large-scale PM2. 5 sensing systems. IEEE Internet Things J. **5**(2), 559–570 (2017)
15. Guo, Y., Liao, W., Wang, Q., et al.: Multidimensional time series anomaly detection: a GRU-based gaussian mixture variational autoencoder approach. In: Asian Conference on Machine Learning, pp. 97–112. PMLR (2018)
16. Tang, C., Xu, L., Yang, B., et al.: GRU-based interpretable multivariate time series anomaly detection in industrial control system. Comput. Secur. 103094 (2023)
17. Zong, B., et al.: Deep autoencoding Gaussian mixture model for unsupervised anomaly detection. In: International Conference on Learning Representations, February 2018
18. Park, D., Hoshi, Y., Kemp, C.C.: A multimodal anomaly detector for robot-assisted feeding using an LSTM-based variational autoencoder. IEEE Robot. Autom. Lett. **3**(3), 1544–1551 (2018)
19. Li, D., Chen, D., Jin, B., Shi, L., Goh, J., Ng, S.-K.: MAD-GAN: multivariate anomaly detection for time series data with generative adversarial networks. In: Tetko, I.V., Kůrková, V., Karpov, P., Theis, F. (eds.) ICANN 2019. LNCS, vol. 11730, pp. 703–716. Springer, Cham (2019). https://doi.org/10.1007/978-3-030-30490-4_56

Research on Script-Based Software Component Development

Yingnan Zhao[1], Yang Sun[1], Bin Fan[2(✉)], and Dan Lu[1]

[1] College of Computer Science and Technology, Harbin Engineering University,
Harbin 150009, China
{zhaoyingnan,ludan}@hrbeu.edu.cn
[2] Harbin Institute of Technology Software Engineering Co., Ltd., Harbin, China
fbina6@qq.com

Abstract. As software demand proliferates and software size and complexity increase, traditional software development models face enormous challenges. As a result, new software development techniques are being explored to meet the requirements of software development. The development of new software development techniques has begun to meet the requirements of software development. Based on many years of experience in object-oriented and component-oriented software design methods, we propose an innovative four-factor scripted analysis and design method for state transfer using the knowledge of computer compilation principles on finite state machine (FSM) circuit module automation. It provides a methodological and empirical reference for programmers in the design and development of application systems using dynamic languages.

Keywords: Software Reuse · Components · Component models

1 Software Component Technology

1.1 Component Technology

Component technology is based on, and well developed by, object-oriented technology. Its aim is to encapsulate objects, including their user interfaces, external interfaces and other properties, as well as the functional implementation of the object into a standardized, standardized whole that can be easily manipulated and used by the component container, making it a common, efficient software component. Component technology is a core technology that supports software reuse and has become a rapidly growing and highly regarded branch of the discipline in recent years [1].

Currently, there is no unified definition of the concept of building blocks, but one of the more widely accepted concepts is that software building blocks are functional units that can be developed, acquired and distributed independently and can interact to form a functional system.

H. Jin et al. (Eds.): GPC 2023, LNCS 14503, pp. 101–112, 2024.
https://doi.org/10.1007/978-981-99-9893-7_8

Normally, the components should have the following three basic properties.

1. Reusability: Reusability is the reason for the existence of building blocks and is the purpose and driving force behind the development of building block technology.
2. Capsulizability: A component is a self-contained entity that encapsulates the content of a design implementation and interacts with the outside world only through an interface.
3. Compositability: Components can be assembled to form a larger whole, and assembly is a means of achieving reuse.

Software Component Technology is based on and well developed by Object-Oriented Technology. It aims to encapsulate objects, including their user interfaces, properties such as external interfaces and the functional implementation of objects into a standardized, standardized whole that can be easily manipulated and used by the component container, making it a common, efficient software component [2].

Software component technology is a core technology that supports software reuse and has become a rapidly developing and highly valued branch of the discipline in recent years [3]. As research into software component technology has progressed, software development methods have changed significantly, and software development methods based on "component-architecture" reuse have become one of the focal points of the software engineering discipline today. This approach is dedicated to the construction of software systems by assembling existing software components, which can be reused at each stage of the development process for new software.

As research into software component technology has advanced and evolved, software development methods have changed significantly, and software development methods based on "component-frame" reuse have become one of the focal points of the software engineering discipline today. This approach is dedicated to the construction of software systems by assembling existing software components, which can be reused at each stage of the development process for new software [4].

1.2   Software Architecture

A software architecture is the organization of the building blocks of a program or system, the relationships between them and the principles and guidelines that govern the design and evolution of the system [5]. In general, the software architecture of a system describes all the computing artifacts in that system, the interactions between the artifacts, the connectors and the constraints on how the artifacts and connectors fit together.

The openness of a software system includes the openness of the data, the openness of the functions and the expandability of the system. Whether a good openness is available basically depends on the system model. The integration of a software system means that the various functional subsystems are unified into the same integrated environment by means of consistent information description and processing mechanisms, and good or bad integration also depends on the system model. The efficiency of a software system usually includes the efficiency of the system operation and the efficiency of the application development. Operational efficiency is the spatial and temporal complexity of the system, while application development efficiency refers to the ease of development and execution efficiency. Efficiency is largely dependent on the system model but is also related to the specific implementation of the system [6].

An open system model makes the assembly of subfunctional components easy to implement and inevitably increases the efficiency of application development; assembly and efficiency in turn contribute to better openness. These three complement each other. Openness is the basis for assembly and efficiency, and openness is the only way to achieve assembly and efficiency.

In response to the openness of application software systems, many types of system models have emerged, representing different stages in the development of application software technology and industry. The representative system models in each stage are the data-centric system model, execution-centric system model, object-oriented system model and bus-based system model.

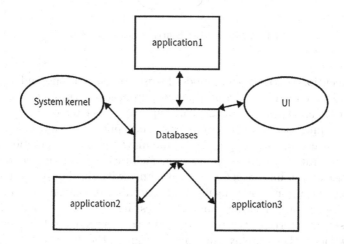

Fig. 1. Data-centric system model

The data-centric system model is shown in Fig. 1. This type of model puts the database at the core level of the system shared, each functional component uses a unified data description, and the development process of each subsystem is completely independent; there is a unified data exchange interface between subsystems; the overall scalability is good, and applications that meet the data exchange standards can be added at will [7]. However, the overall structure of this model is loose and not well assembled; only data reuse can be achieved, not functional reuse, resulting in a large amount of code redundancy. Due to the existence of application-related data, it is difficult to define a data interface standard that meets the needs of all applications, and therefore, data semantic distortion can occur. From the point of view of openness, these systems are only data open, not functionally open, but they are very expandable.

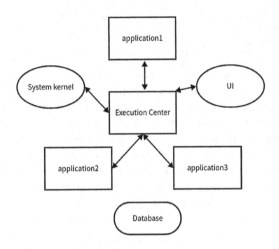

Fig. 2. Execution-centric system model

The execution-centric system model focuses on the sharing and consistency of data and user interfaces across different applications through a unified execution center, as shown in Fig. 2. This type of model separates the common computation and execution functions from the application and places them in the execution center, avoiding code redundancy; the user interaction with the system is separated from the application, facilitating the implementation of a uniform style of user interface; any data exchange with the database is carried out through the execution center, facilitating strict data management and ensuring data consistency [8]. This type of model solves the problems of code redundancy and inconsistent interface styles of data-centric system models, but there are still some drawbacks: the complex functional design of the execution center makes it difficult to define exactly the set of functions that meet the requirements of all applications, and it is also quite difficult to implement; the execution center maintains communication with the user interface and all applications at the same time and manages the data, which is overburdened and prone to bottlenecks. In summary, these models are both open to data and open to functionality and are scalable and overall superior to data-centric system models.

The gradual maturity of object-oriented and middleware technology has introduced new ideas for the modeling of application software systems, resulting in a bus-based system model, as shown in Fig. 3. Components and middleware are new software engineering technologies developed after object-oriented technologies and are an extension of object-oriented technologies. The bus-based system

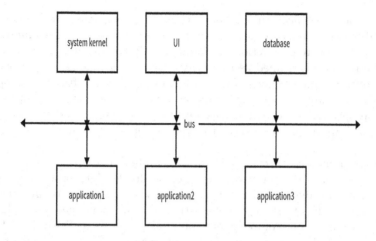

Fig. 3. Bus-based system model

model is still an object-oriented structure, but the objects in the system are modules designed according to specifications. These well-defined software modules (building blocks) coexist in the system and interact fully with each other. Following this structure, several building blocks can be combined to create larger and more complex systems [9].

The key to this model is an efficient bus structure that allows the building blocks to be interconnected with each other with a common interface, enabling plug-and-play and seamless integration of the building blocks. In this model, the number of communication links between the components is linear, and the complexity of communication is greatly reduced due to the consistency of the interface specifications of the components, which also improves the interoperability of the components. The popular CORBA, COM/DCOM and EJB models are all based on this architecture.

One of the two agreements reached at the International Conference on Component-Based Software Development (CBD) held in Kyoto, Japan, in April 1998 was that components cannot be separated from the architecture. Many experts believe that the degree of reuse of 'plug-in' components is directly related to the extent to which they rely on a predefined set of constraints and conventions, which are fulfilled by the architecture.

2   Software Reuse

The concept of software reuse was first introduced in Mcllroy's paper "Mass-produced Software Components" at the 1968 NATO Software Engineering Conference. In the subsequent development process, software reuse technology has made some progress, and there are many reuse technology research results and successful reuse practice activities. In recent decades, as the object-oriented approach has been proposed and has gradually become the mainstream technology, providing basic technical support for software reuse, software reuse research has been highly regarded as a practical way to solve the software crisis, improve the efficiency of software production and improve software quality.

Software reuse is defined as the process of reusing functionally identical or similar software elements and software modules in different software development processes. In a broader sense, software reuse is the development of components at an appropriate level of granularity and their reuse to extend the "system of components" from code-level reuse to requirements analysis models, design and testing.

Software reuse can be examined from a number of perspectives. Depending on the object of reuse, software reuse can be divided into product reuse and process reuse. Product reuse is currently a realistic and mainstream approach. Product reuse refers to the reuse of existing software components and the integration/assembly of the components to obtain a new system. The object of product reuse is reusable components, i.e., components with relatively independent functionality and reusable value.

With reference to the level of abstraction, software reuse can be divided into the following categories.

2.1 Code Reuse

Code reuse, including target code reuse and source code reuse, is one of the most common forms of software reuse. For target code reuse, the runtime environments of most programming languages support such reuse, and this is achieved by providing features such as linking and binding. The source code is reused at a higher level than the target code. When a programmer copies some of the code he or she wants to reuse into a program, it is often compiled incorrectly because the old and new code do not match. To reuse source code on a large scale, a library of reusable building blocks is needed. The library supports the definition of building blocks at the source level to create new systems, and at the target code level, these building blocks can be considered independent reusable building blocks that can be easily assembled and deployed during new system development and flexibly reassembled at runtime [10].

2.2 Design Reuse

The design abstraction level is higher than the source program, so the implementation environment has little impact on the reuse of the design, and reusable components have more opportunities to be reused. There are three ways of doing this: extracting some reusable design components from an existing system and applying these extracted components in the design of a new system; applying the same design to multiple concrete implementations, i.e., reimplementing the entire design of an existing system on a new hardware and software platform; and developing some reusable design components that are not associated with any specific application.

2.3   Analysis Reuse

The level of reuse of analysis is higher than the design outcome. Reusable analysis components are a higher level of abstraction for things and problems in the problem domain, where design techniques and implementation conditions have less influence and therefore have a greater chance of being reused. There are three ways to achieve reuse: extracting reusable analysis components from existing systems and applying these extracted components in the design of new systems; taking a complete analysis document as input and making it possible to implement a number of different designs for different platforms and implementation conditions; and developing reusable analysis components that are not related to any specific application.

2.4   Reuse of Test Information

The reuse of test information consists of two main components: the reuse of test cases and the reuse of test processes. The reuse of test cases is the use of existing test cases for testing new software or the use of previous test cases in a new version of the software after modification. The reuse of test processes is the application of process information recorded during testing in new software testing, which is automatically recorded by software tools. The level of reuse of test information is not easily comparable to the level of reuse of the three categories above because test information is not the same kind of information as the three categories above, nor is it a different level of abstraction of the same thing; it is a different kind of information.

3   Domain Engineering

Software development based on building blocks consists of two subprocesses: domain engineering and application engineering, which are concurrent. Domain engineering and application engineering are both related and distinct: application engineering is the process of developing software for a single application system, i.e., the process of component-oriented software development. The purpose of domain engineering is to create and extract a set of software components that can be easily reused by other software developers and that cross the 'boundary' between domain engineering and the component-based software development process, as shown in Fig. 4.

The aim of domain engineering is to construct reusable components for the development of new systems related to a domain through efficient methods and techniques. The process of domain engineering accomplishes the following three goals: to produce analysis components that represent the domain requirements model, to produce design components that represent the system architecture in the domain, and to produce physical implementations of both of these components, i.e., physical components; application engineering uses the results of domain engineering as a basis for the design of new software systems, and in the development of new software, the requirements analysis and system architecture design phases can use the domain. In the process of new software development, the requirements analysis and system architecture design can be based on the domain model and structural model of domain engineering as a reference, and the analysis and design components of domain engineering can be reused as much as possible. These components are extracted from the component library, certified, adapted and then assembled into new applications. The ultimate goal of domain engineering is to improve the efficiency and quality of software development and to reduce costs by reusing the results of domain engineering to develop applications that meet the needs of specific users in the domain.

Domain engineering and application engineering interact, with application engineering using products made in domain engineering activities to speed up development and application engineering activities influencing future domain engineering development; the two are iterative and together constitute reuse engineering. Individual systems are the focus of application engineering, and the manufacture of one or more related systems within a domain is the focus of domain engineering, which supports application. Engineering a series of individual systems by designing a system solution based on a system relationship tree.

3.1   The Three Basic Stages of Domain Engineering

The ultimate goal of domain engineering is to achieve results that improve development efficiency and quality and reduce development costs when engineering applications that meet specific user needs by reusing the results, such as architectures and artifacts. The core of the domain engineering outcome is the creation and refinement of a domain-oriented library of reusable architectures and artifacts, the whole process being iterative [11].

Domain engineering consists of three phases: domain analysis, domain design and domain implementation.

(1) Domain analysis: The goal of this phase is to obtain a domain model.

Prior to this, some preparatory work is necessary, such as the definition of domain boundaries, the definition of the objects to be analyzed, and the identification of information sources. Once these preparations have been completed, the system requirements in the domain are analyzed and classified, those that are widely shared are identified and extracted, and the domain model is built. The scope of domain requirements is constantly changing, and domain experts determine the scope of domain requirements by examining sample system requirements. In contrast to systems analysis, domain analysis focuses on a class of identical or similar applications, and once a good domain requirements model is obtained, it is used repeatedly in future system development.

(2) Domain Design: The goal of this phase is to obtain a domain-specific software architecture (DSSA), which represents a high-level design that can accommodate the requirements of multiple systems in the domain rather than just a single system and describes the solution to the requirements expressed in the domain model. The solution [?]

DSSA represents the common parts of the software architecture of systems in the domain and is an important object for the implementation of change binding in domain engineering. The domain model is the key component that describes the common parts of the requirements, which are obtained by examining and summarizing the requirements descriptions of existing systems.

(3) Domain implementation: Development and organization of reusable information based on the domain model and DSSA.

This reusable information may already exist in a repository of components or be extracted from an already existing system, or it may be newly developed. The timing of reuse of this information is determined by the domain model and DSSA, which therefore support systematic software reuse.

Domain engineering is an iterative, iterative, progressive refinement process, where it is possible to go back to previous steps and modify and refine them, then return to the current step, and so on and so forth at each stage of the domain engineering implementation.

4   Our Approach

First, to analyze the four elements of each link for the business logic of the application system, we must find the state of each link on its behalf. Here, it is recommended to use the UML user case diagram for analysis because the user case diagram is the most capable of dividing the system operator and its operation object into a system analysis diagram.

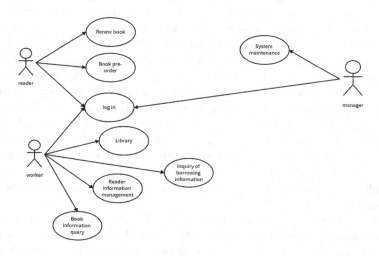

Fig. 4. User case diagram in state analysis

Next, we need to find the conditions and actions in the four elements through the state, which requires the reduction of the business logic structure, and to find the relationship between the states. We can use the object flow diagram of the object and message request method to describe the business logic on the basis of Fig. 4, as shown in Fig. 5.

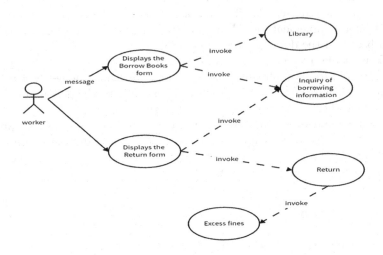

Fig. 5. State transition relationships

The diagram takes the design of staff to borrow and return books as an example. Staff first enter the service request state. The state - generally in the system is mostly a functional navigation form state. In this state, according to the different requests of the user to open and display the borrow or return books from the state, in the above form state after the user enters or selects the specified book and reader information, the user will issue the borrow sure or return sure. In the above form state, after the user enters or selects the specified book and reader information, the user issues the operation instruction of borrowing or returning the book to perform the operation of borrowing and returning the book. Among them, during the operation of returning the book, the operation of overdue fine check will also be performed, and if there is no overdue, the operation of returning the book will be completed and the operation of returning the book will be completed. However, if the book return time is found to be overdue, the book will be put into the overdue fine waiting state v waiting for the user's processing instructions.

Here, we use UML state diagrams to first list the states and then connect the relationships between the states with directed arrows, define the transfer conditions and transfer actions at the arrows, and use the [transition conditions]/[transition actions] structure to describe them, as shown in Fig. 6.

In the diagram, we have scripted a business logic subprocess on book lending using a four-factor structured analysis of state transfer. The scripted design clearly reflects the state, conditions and actions of the process using scripts, e.g., a staff member unconditionally opens a navigation form state called "NaviForm"

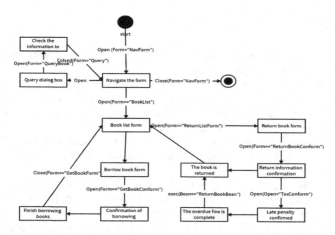

Fig. 6. Scripted design of application system business logic

from the initial system entry state. This state can then perform an "Open" action to open the form based on the condition that a menu item command of type "MenuSelect" with a value of "BookList" is satisfied. This action determines whether the user has permission, and if so, creates and initializes a form object called "BookListForm" and displays it; for example, when a book is returned and the book lending time is found to be overdue, the overdue fine information is issued to confirm the status, and when the user submits a menu item called "OK", the system will perform the action of paying the fine to complete the operation of overdue fines. The action is "exec" and is performed by a state transfer method "Tax". In addition, so on, we will have a complete description of the scripted result of the business logic of an application in this state diagram.

5   Conclusion

The result of scripting the business logic of an application system is only the first step in the rapid development of an application system using a dynamic language, but just as the success of a movie must come from a good script, how to write a "script" and how to write a good "script" will have a bearing on the ultimate success or failure of the rapid development of the scripted application system.

Acknowledgement. This work was supported by the National Key R&D Program of China under Grant No. 2020YFB1710200.

References

1. Fuqing, Y., Hong, M., Keqin, L.: Software reuse and software component technology. Acta Electronica Sinica (1999)
2. Jang, C., Song, B., Jung, S., et al.: A development of software component framework for robotic services. In: Fourth International Conference on Computer Sciences & Convergence Information Technology. IEEE Computer Society (2009). https://doi.org/10.1109/ICCIT.2009.161
3. Kohn, Dykman, Kumfert, et al.: Scientific Software Component Technology. Office of entific & Technical Information Technical Reports, 2000(9). https://doi.org/10.2172/792560
4. Kim, S.D., Her, J.S., Chang, S.H.: A theoretical foundation of variability in component-based development. Inf. Softw. Technol. **47**(10), 663–673 (2005). https://doi.org/10.1016/j.infsof.2004.11.007
5. Jingga, K., Sunindyo, W.D.: Component-based development using moodle as alternative for E-learning software development. In: International Conference on Information Technology and Electrical Engineering (2020). https://doi.org/10.1109/ICITEE49829.2020.9271670
6. Ali, M.A., Ng, K.Y.: Software component quality model. Int. J. Eng. Adv. Technol. **9**(1), 1758–1762 (2019). https://doi.org/10.35940/ijeat.A2659.109119
7. Zhou, K., Zhu, X.: Development of distributed component technologies. Wavelet Analysis and Active Media Technology vol 1. Chongqing Automobile Institute, Chongqing 400050, China (2005). https://doi.org/10.1142/9789812701695_0051
8. Crnkovic, I.: Component-based software engineering - new challenges in software development. Softw. Focus **2**, 127–133 (2001). https://doi.org/10.1002/swf.45
9. Hribar, L., Huljeni, D.: Modeling non-functional characteristics of signaling protocols software components. In: 2015 International Conference on Advanced Educational Technology and Information Engineering(AETIE 2015).0[2023-07-01]. ConferenceArticle/5af2b05fc095d70f18a130d5
10. Li, Y., Jiang, L.: Method of software development based on component technology. Adv. Intell. Syst. Res. (2012). https://doi.org/10.2991/citcs.2012.192
11. Abmann, U.: Software Architectures and Component Technology. [2023-07-01]

Integration Model of Deep Forgery Video Detection Based on rPPG and Spatiotemporal Signal

Lujia Yang, Wenye Shu, Yongjia Wang, and Zhichao Lian[✉]

School of Cyber Science and Engineering, Nanjing University of Science and Technology,
Nanjing 210000, China
lzcts@163.com

Abstract. With the development of deep learning, video forgery technology is becoming more and more mature, which may bring security risk and the further development of forgery detection is urgently needed. Most of the existing forgery detection technique are based on artifacts and detail features, which are greatly affected by the resolution, and its generalization ability needs to be improved. In this paper, a multi-modal fusion forgery detection model architecture based on the inherent biological signals and spatio-temporal signals in videos is proposed. In the process of forgery detection, the model first recognizes the face of the video. Subsequently, video frame extraction and rPPG signal extraction based on Green channel are performed on the video, respectively. These two data are later input into 3D and 2D convolutional neural networks to train the base learner respectively. Finally, the integration model is constructed based on stacking strategy. Sufficient experiments show that the established fusion model can cope well with low-resolution cases and has good generalization performance, achieving 93.38% and 91.57% accuracy on FF++ c23 and celeb-DF-v2 data set, respectively.

Keywords: Deep Fakes · Fake Detection · Biological Signal · Spatiotemporal Signal · Deep Learning · Ensemble Learning

1 Introduction

In recent years, with the rapid development of computility, kinds of models and algorithms, deep learning technology has developed by leaps and bounds and made major breakthroughs in the field of computer vision. It is now widely used in image classification, target recognition, image generation and other difficulty tasks. On the one hand, such powerful technology has promoted progress in many fields. On the other hand, it also leads to safety hazard.

Since "Deepfakes" [1], the user, post fake porn videos of celebrities and caused an uproar on social network in 2017, deep forgery techniques on video has emerged in succession. The past few years have witnessed the advent of mass-market and recreational face-changing apps, such as Zao [2], Face APP [3], FakeAPP [4], Faceswap [5] and etc., which further lower the technical threshold of making fake videos. Worries have multiplied about deep forgery technology being intentio-nally misused. The importance of the research of video deep forgery detection technology is self-evident as the threat of the abuse of deep forgery technique is showing up.

The existing deep learning video forgery technologies mainly focus on face forgery. According to the difference of principles and algorithms to be used, it can be divided into forgery based on traditional computer graphics and forgery based on deep learning. Application of the former often has certain requirements for the 3D similarity between the fake face and the real face, and its forgery results have low resolution as well as poor effect, which can be easily identified by the artifacts in the fake video. While the forgery technology based on deep learning, especially a series of forgery methods based on generative adversarial networks (GAN) [6], such as ProGAN [7], StyleGAN [8], StarGAN [9], etc., can not only generate images with high visual quality, but also realize multi-style forgery tasks, which greatly increase the difficulty of forgery detection.

Nowadays, video forgery detection models are mainly based on the difference of color and other feature texture between the generated and natural images to identify the authenticity of the video. However, it doesn't always work. When the fake videos are transmitted multiple times through different streaming media on the network, its resolution will be degraded, resulting in the disappearance of artifacts, and the features in the videos will be disturbed by noise, which makes it difficult to distinguish.

In this paper, biometric features inherent in videos that are difficulty to be forged are utilized. The green channel com-ponents method is used to extract rPPG (remote photople-thysmography) features, a biological signal related to heart rate, and the convolutional network is used to learn its spatial coherence and temporal consistency for forge detection. Additionally, spatio-temporal signals extracted and learned independently by three-dimensional convolutional network are introduced into the model to further improve the accuracy of forgery detection. An integrated model based on the fusion of biological signals and spatio-temporal signals is constructed.

The main contributions of our paper are as follows:

- Different modes of information, i.e. temporal signal, spatial signal and biological signal are simultaneously introduced into the model for forgery detection,
- A Multi-modal integrated forgery detection model based on Stacking, combining 2D and 3D convolutional neural networks, is proposed,
- Experiments show that the integrated model in this paper is less affected by the resolution, and has strong generalization ability of detection to various forgery algorithms.

2 Related Work

In this section, deepfake methods, detection methods, and deepfake detection datasets are introduced.

2.1 Deep Fake Methods

Up till now, the face deepfake methods can be roughly divided into four types: identity replacement, face reenactment, attribute editing and face generation.

Identity replacement refers to the whole or partial replacement of the original face. The global face replacement algorithms include Deepfakes [1], FaceShifter [10], etc. The partial replacement algorithms include RAFSwap [11], and the model preserving appearance attributes such as illumination and skin color proposed by Xu et al. [12].

Face reenactment refers to the transformation of facial expression or head pose. The deferred Neural Rendering model proposed by Thies et al. [13], and the X2Face model proposed by Wiles et al. [14] and other models integrate deep learning into the original 3D modeling method to reduce artifacts and speed up face changing.

Attribute editing refers to the modification of a person's facial attributes such as hair color, age, and whether he or she has accessories such as glasses. The early property editing methods inevitably affected other properties when they changed one property. After the emergence of deep learning, Schwarz et al. [15] proposed generative radiance field (GRAF), and Shen et al. [16] proposed semantic decoupling based on pre-training generation techniques. The controllability of att-ribute editing is achieved by decoupling the attributes.

Face generation refers to the generation of a new face from noisy information that does not exist in reality. Representative algorithms include Pro-GAN [7], StyleGAN [8] and others based on generation technology.

Forgery algorithms based on deep learning techniques, especially generative techniques, have high forgery efficiency, high quality and few artifacts. However, there are still differen-ces between the fake video and the real video, such as detailed texture, color features, temporal and spatial consistency.

2.2 Deep Fake Detectors

Most of the existing fake videos are made by splicing the forged frames of the real video. Therefore, it can authenticate the video by frame forensics, or by checking the consistency of consecutive video frames.

Current forgery detection techniques can be divided into image forensics based methods and continuous video frames based methods according to the different information modes used. And the former can be further divided into forgery detection on spatial domain and frequency domain. Spatial information mainly refers to the pixel information in each frame of the video. Forgery detection based on spatial signals mainly focuses on finding artifacts, GAN fingerprints and other abnormal detailed texture features in fake video frames. Representative models include Xception [17], EfficientNet [18], Face X-Ray [19], etc. The spatial signal used by this kind of method is easily affected by the resolution of the image. When facing the untrained forgery method, its accuracy will be reduced, and the generalization of the model is limited.

The frequency domain information refers to the characteristics of the video signal in the frequency domain, including frequency and amplitude distribution. The forgery detection method based on time domain signal tries to mine the details of artifacts in the frequency domain for forgery detection. For example, Qian et al. [20] constructed a frequency domain perception module to adaptively extract frequency domain forgery clues for authentication. Li et al. [21] obtained forgery artifacts from different frequency bands by constructing frequency domain feature generation modules. Based on the Transformer model, Wang et al. [22] realize the features extraction of the local artifact in different levels of space, and integrate frequency domain features for deepfake detection. This method can maintain good performance on highly compressed images, but its generalization is still poor.

The methods of video forgery detection using continuous information mainly use the inconsistency of time domain information. According to the difference of the consistency criteria used, they can be divided into consistency checking methods based on biological signals, time series, and human behavior. Conotter et al. [23] extracted color jitter from the forehead of a human face by 3D face tracking method to obtain pulse related signals for authentication. However, this method is greatly affected by resolution, illumination and occlusion, and the accuracy of the signal is low. Li et al. [24] proposed to detect authenticity by extracting the blink frequency. Hu et al. [25] detected the false by judging the consistency of the bright surface of the eye cornea. Amerini et al. [26] proposed the optical flow vector field to obtain the timing change and they using VGG-16 to judge the rationality of the change to identify fake videos. Mittal et al. [27] performed an analysis based on the emotional characteristics of the characters contained in the video and audio. Most of these methods use specific signals for forgery detection, and their accuracy will decline in the face of adversarial attack technology.

This paper makes full use of all kinds of signal in video, such as biosignals on time domain, as well as signals on spatial and frequency domain and integrates the multi-modal fusion forgery detection model. The accuracy of forgery detection of our model is less affected by the resolution of the picture, and the adaptability of the restoration is good.

3   Method

This section will introduce the fusion model constructed in this paper and various methods used in the model.

3.1   Integration Model

The information in video can be divided into frequency domain information, spatial domain information and time domain information. The frequency domain and spatial domain information is the video detail texture, artifact features and frequency details obtained from a single frame of the video. Temporal information refers to dynamically changing information obtained from consecutive frames, such as facial expression, eye gaze, mouth movement and other biological signals. The fake video generated by the existing forgery technology is different from the real video in frequency domain, spatial domain and time domain information. Therefore, this information can be extracted from the video and fed into a specific model for forgery discrimination.

However, the forgery detection method based on a certain modality has some defects in robustness and generalization. For example, although the forgery detection method based on spatial information can capture the local and global color and texture features of the video image frame, it is easily affected by video compression rate and forgery method. The forgery detection method based on frequency domain information can obtain the

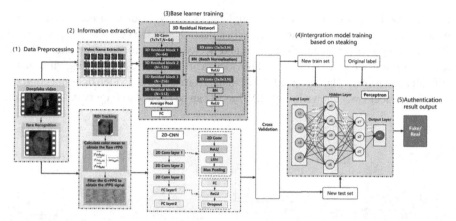

Fig. 1. Architecture of fusion forgery detection model. In forgery detection, the model first performs face recognition (1), and extracts video frames and rPPG signals (2). Then respectively input them into 3D and 2D convolutional neural networks to train the base learners (3). Based on Stacking, the base learners are cross-validated to obtain the training and test data sets of ensemble learning and to train the final intergration model (4).

subtle frequency domain changes in the video, while loses more spatial domain information such as texture, which easily leads to the degradation of classification performance. The forgery method based on time domain can extract time series features, but it cannot use the frequency domain and spatial domain information in the video frame, and the forgery detection effect is affected by the used time domain signal type.

In this paper, the time-domain biological signals in the video that are difficult to be forged by advanced forgery technologies (GAN, etc.) are selected as the basis for forgery detection. The 2D-CNN is used to learn the difference between the rPPG signals in the real and fake videos. A 3D residual network is added to the model to learn the spatial domain, frequency domain signals and other time-domain signals in the video. And a integration model is constructed to make full use of the data of each mode in the video to carry out multi-modal forgery detection.

The architecture of the fusion model is shown in Fig. 1. The specific steps of the model forgery detection are as follows: (1) **Data preprocessing**: perform face recognition based on Retinaface [28] on the video, lock the face, and crop the video. (2)–(3) **Information extraction & base leaner training**: a) Video frames are extracted from the video and input into the 3D residual network to train the 3D base learner. b) rPPG signals are extracted from the region of interest of the face based on the Green channel method and input into the 2D convolutional network to train the 2D base learner. (4) **Intergration model training based on Steaking**: integrate the base learners to train the fusion model and obtain the final identification model. (5) **Output authentication result.**

3.2 rPPG

Remote Photoplethysmography (rPPG) signal [29] is a biological signal related to heart rate which can be obtained from video. In the human body, the heart beats regularly and the blood is pumped regularly throughout the body, leading to the blood vessels expanding and contracting accordingly. When the light hits the skin, the camera receives the light component related to the heart rate reflected back from the subcutaneous blood vessels, forming a regular rPPG signal in video. This tiny but regular biological signal is difficult to be forged. The extracted rPPG signal from the forged video does not conform to the biological law. (Shown in Fig. 2.) Thus, it can be applied to forgery detection [30].

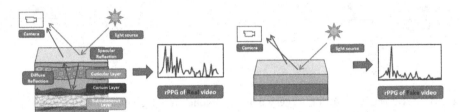

Fig. 2. Schematic comparison of real and fake video rPPG signals

There are many methods to extract rPPG signals from videos. In this paper, Green-channel-based PPG signals (G-rPPG), which are less affected by image compression, are used as the detection basis.The rppg signal formula in the optical model is constructed based on the absorption and scattering characteri-stics of skin tissue to light as formula (1).

$$C_k(t) = I(t)(V_s(t) + V_d(t)) + V_n(t) \qquad (1)$$

The pixel value of skin k at time t of video is $C_k(t)$. It consists of light intensity $I(t)$, skin specular reflection comp-onent $V_s(t)$, skin diffuse reflection component $V_d(t)$ and noise $V_n(t)$. The light intensity $I(t)$ and skin specular reflection $V_s(t)$ are mainly related to the fixed brightness of the light source, the light Angle and the light distance. The diffuse component $V_d(t)$ is mainly affected by skin color and pulse. Among the three RGB channels, the diffuse component $V_d(t)$ accounts for the largest proportion in the Green channel. After normali-zation and filtering, most constant that does not change with time and meaningless noise are removed, and the Green channel signal can be approximately regarded as the rPPG signal.

The detection model based on rPPG signal uses biological signals on time domain in the video. It works by detecting whether the temporal signals in the same face area match the peaks and troughs of the human heart rate (i.e. Time consistency), and the similarity of rPPG signals at different locations (i.e., spatial coherence) to determine video authen-ticity. The detection method is less affected by the forgery method and video resolution, resulting in good generalization.

3.3 R3D

3D Residual Network (R3D) is an extended network of ResNet. Through the introduction of three-dimensional convolution kernel in ResNet network, the model can learn the time feature from the continuous frames of the video and capture the timing information, the relationship between frames and the action of the video. The introduction of three-dimensional pooling layer enables the model to extract spatial features from the video as well as capture the relationship between pixels in the video frame, realizing the convolution operation in space-time dimension.

The process of deepfake detection and the construction of the network of R3D are shown in Fig. 3.

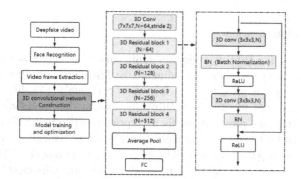

Fig. 3. Detection flow and network structure of R3D

When detecting the video, R3D divides the video into several short video clips, and preprocesses the clips by cropping, alignment and labeling,etc. The video is then decomposed into several consecutive video frames, which will be later input into 3D Residual network for model training and video detection. 3D Residual network normally consists of convolutional layers, 3D residual blocks, pooling layers, and fully connected layers (FC). The 3D residual block consists of multiple convolutional layers, batch normalization layers (BN), skip connections, and activation function layers. Figure 3, illustrates the network structure of 3D ResNet 18 (i.e.model with a network depth of 18). The input of the network is a 3D tensor, which contains the time series information of the video and the spatial information of each frame. The first layer of the network, the 3D convolution, captures the relationship between temporal information and spatial information, as well as fuse the two information together. In the subsequent layers, the network gradually extracted high-level features by stacking multiple residual blocks, and obtained the classific-ation results through average pooling and fully connected layer.

3.4 Stacking

Model fusion is to fuse multiple training models into one more complex model with better generalization ability accord-ing to a certain method strategy. Stacking and weighted fusion are two common ideas for fusion of different models. Nevertheless, the weight of weighted fusion will greatly affect the accuracy of the model while Stacking are more flexible, and can combine different models according to specific needs, so as to achieve better results.

The specific steps of Stacking model fusion are as follows: 1) *Dataset Split*: Split the dataset into training set and test set by k-fold cross-validation. 2) *Base model training*:

Multiple base models are trained with the training set from the first part. 3) *Generate predictions:* Obtain the prediction results of each metamodel on the test set. 4) *Metamodel training*: Using the label of the test set and the corres-ponding base model prediction results as the train set to train a metamodel. After training, the weights of each base classifier in the ensemble model can be obtained. 5) *Generate final predictions*: Input the prediction results of each base learner for the target into the trained metamodel to generate the final prediction results.

3.5 Bayesian Optimization

Bayesian Optimization (BO) is a global optimization algorithm based on probability distribution. The key parts of the algorithm are the Prior Function (PF) and the Acquisition Function (AF). A prior function is used to compute the likelihood of a model parameter and treat it as an actual probability distribution, thus modeling the parameter space. Among them, the most commonly used prior function is the Gaussian process, which can build prior knowledge of the parameter space and calculate the likelihood of individual parameter points, so as to achieve consistent optimization. The acquisition function is a tool for actual optimization, which can calculate the most likely parameter points in the parameter space according to the prior function model, and obtain the optimal parameter points through actual testing.

Pseudocode for Bayesian optimization based on Gaussian Process is as follows:

Algorithm：Bayesian Optimization

Input：T, f, M

Output：The optimal combination of hyperparameters

1：**function** *Bayesian Optimization*(T, f, M)
2：   Initialize the parameter space P and dataset D
3：   Construct a initial Gaussian process model $GP(D, M)$
4：   **for** $t = 1$ to T **do**
5：   Find x_t by optimizing the acquisition function over the GP:
$$x_t = argmax_{x \in P} EI(x|f, \varsigma_t - 1)$$
6：   Sample the objective function: $y_t \leftarrow f(x_t) + \varepsilon_t$
7：   Augment the data $D_{1:t} \leftarrow D_{1:t-1} \cup (x_t, y_t)$
8：   Update the$GP(D, f)$ and get the posterior distribution $f_t(x|D_t)$
9：   **end for**
10：   **return** $x \leftarrow argmax_{x \in P} f_t(x|D_t)$
11： **end function**

---

f is the objective function to be optimized. M is the prior distribution function. Gaussian process (GP) is the surrogate function for the objective function f. D is the data set. EI is the Acquisition function. Bayesian Optimization builds a surrogate model to fit the objective function and uses Bayesian methods to update the posterior distribution of the objective function to get the optimal combination of hyperparameters.

Bayesian optimization algorithm can adaptively explore the search space, so as to quickly find the global optimal solution and avoid falling into the global optimum resulting in model underfitting.

4   Experiment

In this section, the detection performance of the model constructed in this paper is test and compared with other recent deepfake detection algorithms. This paper also tests the effect of the model in the face of fake videos with different compression rates produced by different forgery methods.

4.1   Datasets

Here are some commonly used deep face forgery detection datasets: Faceforensics++ [31], DEDCP [32], Celeb-DF-v2 [33], etc.

The **FaceForensics++** [31] dataset contains videos with two different compression ratios (i.e. FF++ c23 and FF++ c40). Each includes 1000 original Youtube video and 5000 fake video processed by Face2Face [34], FaceSwap [5], DeepFakes [1] and Neural-Textures [24]. With a large amount of video data and multiple types of deepfake technologies being used, FaceForensics++ is one of most used data set for testing deepfake detection technologies. Yet, there is still room for improvement in the falsification of this dataset, and the facial artifacts of fake video in it are relatively obvious.

Celeb-DF-v2 [33] is a face forgery detection dataset created by Li et al.. It contains 590 real videos and 5639 fake videos based on improved deepfake technology. The forged videos in the dataset are of high quality with few artifacts, making it a more challenging dataset.

In this paper, the FaceForensics++ and Celb-DF-v2 datasets are selected as the training and test sets to compare the performance of the model with other algorithms. And on the basis of the original FF++ dataset, this paper adds the fake videos generated by the FaceShifter [10] method from the original FF++ real videos with compression rates of 23 and 40 respectively to the datasets. Accuracy rate of detection (Acc), False Positive Rate(FPR) and Area Under ROC curve (AUC) are selected as the evaluation criteria for model testing.

4.2   Comparison

Table 1 compares the performance of the proposed model in this paper with other forgery detection model algorithms on FaceForensics++ and Celeb-DF-v2 datasets.

It can be seen from the table that compared with other authentication algorithms, the model in this paper has a good authentication effect on Celeb-DF-v2 dataset, and can reach an accuracy of 91.57%, which is better than the performance of most current algorithms. This may be because the video resolution in the Celeb-DF-v2 dataset is high, and the proposed model architecture makes it possible to learn more information in high-resolution videos than other algorithms, resulting in better video classification performance.

Table 1. Detection accuracies Comparison.

Model	FF + raw	FF++ c23	FF++ c40	Celeb-DF-v2
Face X-ray [17]	99.1%	78.4%	34.2%	79.5%
Xception [19]	99.26%	95.73%	81.00%	–
LipForensics [31]	98.9%	98.8%	94.2%	86.9%
MesoNet [32]	95.23%	83.10%	70.47%	54.8%
Steg. Features + SVM [32]	97.63%	70.97%	55.98%	–
Intergration model	–	93.38%	88.77%	91.57%

On the videos of FaceForensics++ dataset with compression rates of 23 and 40, the accuracy of the proposed model can reach 93.38% and 88.77% respectively, which is excellent and can catch up with the current level of excellent algorithms. However, there is still more room for model improvement when comparing it with advanced algorithms, such as Xception [19] and LipForensice [36]. The model in this paper is mainly trained on Celeb-DF-v2, and it is believed that through further parameter adjustment and training, it can show better performance on the FF++ dataset.

Moreover, from the Table 1, we can infer that the deepfake detection accuracy of the intergration model is less affected by the video resolution since it can still maintain more than 88% accuracy on videos with low resolution. Yet, most of the algorithms have a significant decline in the accuracy of forgery detection on low-resolution videos, such as Face X-ray [17] and Steg. Features [37] and so on.

4.3   Generalization Testing

The forgery detection accuracy and missed detection rate(false positive rate, fpr) on Celeb-DF-v2, FF++ c23 and FF++ c40 datasets of the intergration model, which is mainly trained on Celeb-DF-v2, are shown in Table 2 and Fig. 4.

On deepfale videos faked by Deepfake, Face2face, FaceShifter, Faceswap, and NeuralTextures, the interdration model shows good generalization. It has high accuracy and low false positive rate for all kinds of forgery detection methods (Fig. 4), proving that it can adapt to unseen forgery algorithms and low resolution cases.

Nevertheless, the overall accuracy is affected by false negative rate. Due to the imbalance in the proportion of true and fake videos in the FF++ and Celb-DF-v2 datasets, the dataset itself is biased to judge the video as fake, making the model trained on them easily judge the real video as fake. And the accuracy of forgery detection needs to be further refined.

During the tests, this paper found that the intergration model is slightly worse at detecting fake videos from Face2Face and NeuralTextures. It is likely that as a result of different key forgery regions for different forgery methods. These two methods focus on expression forgery, and the important signals for forgery detection most exist around the lips while the region of interest of the proposed model is mainly concentrated on the cheek and forehead. And the deepfake method with the region of interest on the

Table 2. Generalization of integration model on FF++.

Datasets		Acc	FPR	AUC
Celeb-DF-v2		91.57%	0.73%	0.8094
FF++ c23	Deepfake	99.60%	0.40%	–
	Face2face	98.59%	1.41%	–
	Faceshifter	98.79%	1.21%	–
	Faceswap	97.39%	2.61%	–
	NeuralTextures	88.84%	11.16%	–
	all	93.38%	3.36%	0.8686
FF++ c40	Deepfake	99.60%	0.40%	–
	Face2face	98.29%	1.71%	–
	Faceshifter	98.89%	1.11%	–
	Faceswap	98.69%	1.31%	–
	NeuralTextures	96.98%	3.02%	–
	all	88.77%	1.50%	0.7003

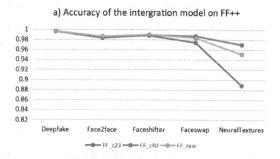

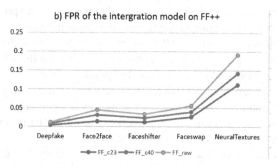

Fig. 4. Generalization of the intergration model on FF++. The intergration model trained on Celeb-DF-v2 can achieve an accuracy of more than 88% and an FPR of less than 20% on fake videos generated by different forgery methods, showing good generalization.

lips performs particularly well on these two fake methods, such as LipForensics. For the method of forging the full face or directly changing the face, such as Faceswap, Deepfake and improved Deepfake (i.e. the method used in Celeb-DF-v2), this paper can better capture the forged information in the video and accurately classify it.

4.4 Ablation

In this section, ablation experiments are carried out for the ensemble model, and it is proved in Table 3 that the ensemble model has better accuracy and AUC both on the FF++ dataset and Celeb-DF-v2 dataset. It is difficult to achieve high accuracy when training r3D or rPPG + 2D-CNN models alone as achieving good accuracy on these two base learners requires careful parameter tuning and thoughtful training. But by constructing the intergration model based on the Stacking and selecting parameters based on the Bayesian optimization algorithm, a better model can be obtained more easier.

Table 3. Results of ablation experiments.

Datasets	FF + + c23		FF + + c40		celeb-DF-v2	
Index	Acc	AUC	Acc	AUC	Acc	AUC
Intergration Model	93.38%	86.86%	88.77%	70.03%	91.57%	0.7094
rPPG + 2D-CNN	81.10%	65.29%	72.80%	55.37%	82.30%	0.6056
r3D	80.70%	78.40%	71.20%	67.00%	86.20%	0.8330

5 Conclusion

This paper proposes a multi-modal fusion deepfake detection model architecture based on the biological signal and spatio-temporal signal in the video, and provides some idea and reference for the construction of the forgery detection model.

The model in this paper extracts rPPG signals from the Green channel of the face ROI region, and uses 2D convolutional neural network to learn the difference between real and fake videos in spatial coherence and temporal consistency of rPPG signals for forgery detection. Meanwhile, the 3D residual network is introduced into the model to learn more information. The temporal features in the video are learned by the 3D convolution kernel, and the spatial features are extracted by the pooling layer. These learned features are used to assist video classification.

This model makes good use of the time domain, frequency domain and spatial domain information in the video. The experimental results show that the model has good performance. It can achieve 93.38% accuracy on FF++ c23 and 91.57% accuracy on Celeb-DF-v2 datasets, which is better than most forgery detection models. In the face of low-resolution video, the forgery detection accuracy can be maintained at more than 85%. And in the face of fake videos made by different forgery methods, the accuracy of detection can maintain above 88%, proving that the generalization ability of the model is good.

Nevertheless, this paper mainly focuses on the detection of face deepfake videos.Thus, the performance of our model in other forgery fields is relatively limited. And since it is an integrated model combining 2D and 3D CNN together, the time and space complexity of the model is not ideal to some extent. In future studies, we will pay more attention to the relevant issues and improve our model.

Acknowledgment. This work was supported by the National Key R&D Program of China (2021YFF0602104-2).

References

1. Deepfakes. Deepfakes github [EB/OL]. https://github.com/Deepfakes/facesw-ap. Accessed 08 May 2023
2. Zao: Zaoapp [EB/OL], 01 Dec 2019. https://zaodownload.com/download-zao-app-deepfake. Accessed 08 May 2023
3. Faceapp. Faceapp [EB/OL]. https://apps.apple.com/gb/app/faceapp-ai-faceedi~tor/id1180 884341. Accessed 08 May 2023
4. FakeAPP. FakeAPP [EB/OL]. https://www.fakeapp.org. Accessed 08 May 2023
5. FaceSwap. FaceSwapgithub [EB/OL]. https://github.com/MarekKowalski/FaceSwap. Accessed 08 May 2023
6. Goodfellow, I.J., et al.: Generative adversarial nets. In: Proceedings of the 27th International Conference on Neural Information Processing Systems, pp. 2672–2680. MIT Press, Montreal, Canada (2014)
7. Karrast, T., Ailat, T., Laines, S., et al.: Progressive growing of GANs for improved quality, stability, andvariation. In: Proceedings of the 6th International Conference on Learning Representations. Vancouver, Canada: Open Revier.net (2018)
8. Karras, T., Laine, S., Aittala, M., Hellsten, J., Lehtinen, J., Aila, T.: Analyzing and improving the image quality of styleGAN. In: Proceedings of 2020 IEEE/CVF Conference on Computer Vision and Pattern Recognition, pp. 8107–8116. IEEE, Seattle, USA (2020). https://doi.org/10.1109/CVPR42600.2020.00813
9. Choi, Y., Uh, Y., Yoo, J., Ha, J.W.: StarGAN v2: diverse image synthesis for multiple domains. In: Proceedings of 2020 IEEE/CVF Conference on Computer Vision and Pattern Recognition (CVPR), pp. 8185–8194. IEEE, Seattle, USA (2020). https://doi.org/10.1109/CVPR42600.2020.00821
10. Li, L.Z., Bao, J.M., Yang, H., et al.: Advancing high fidelity identity swapping for forgery detection. In: Proceedings of 2020 IEEE/CVF Conference on Computer Vision and Pattern Recognition, pp. 5073–5082. IEEE, Seattle, USA (2020)
11. Xu, C., Zhang, J.N., Hua, M., et al.: Region-awareface swapping. In: Proceedings of 2022 IEEE/CVF Conference on Computer Vision and Pattern Recognition, pp. 7622–7631. IEEE, New Orleans, USA (2022)
12. Xu, Y.Y., Deng, B.L., Wang, J.L., et al.: High-resolution face swapping via latent semantics disentanglement. In: Proceedings of 2022 IEEE/CVF Conference on Computer Vision and Pattern Recognition, pp. 7632–7641. IEEE, New Orleans, USA (2022)
13. Thies, J., Zollhöfer, M., Nießner, M.: Deferred neural rendering: Image synthesis using neural textures. ACM Trans. Graph. **38**(4), 66 (2019)

14. Wiles, O., Koepke, A.S., Zisserman, A.: X2Face: A network for controlling face generation using images, audio, and pose codes. In: Ferrari, V., Hebert, M., Sminchisescu, C., Weiss, Y. (eds.) ECCV 2018. LNCS, vol. 11217, pp. 690–706. Springer, Cham (2018). https://doi.org/10.1007/978-3-030-01261-8_41

15. Schwarzk, L.Y.Y., Niemeyer, M., et al.: Graf: generative radiance fields for 3D-aware image synthesis. In: Proceedings of the 34th International Conference on Neural Information Processing Systems, p. 1692. Curran Associates Inc., Vancouver, Canada (2020)

16. Shen, Y.J., Gu, J.J., Tang, X.O., et al.: Interpreting the latent space of GANs for semantic face editing. In: Proceedings of 2020 IEEE/CVF Conference on Computer Vision and Pattern Recognition, pp. 9240–9249. IEEE, Seattle, USA (2020)

17. Chollet, F.: Xception: deep learning with depthwise separable convolutions. In: Proceedings of 2017 IEEE Conference on Computer Vision and Pattern Recognition, pp. 1800–1807. IEEE, Honolulu, USA (2017)

18. Tan, M.X., Le, Q.V.: Efficient net: rethinking model scaling for convolutional neural networks. In: Proceedings of the 36th International Conference on Machine Learning, pp. 6105–6114. PMLR, Long Beach, USA (2019)

19. Li, L.Z., Bao, J.M., Zhang, T., et al.: Face X-ray for more general face forgery detection. In: Proceedings of 2020 IEEE/CVF Conference on Computer Vision and Pattern Recognition, pp. 5000–5009. IEEE, Seattle, USA (2020)

20. Qian, Y., Yin, G., Sheng, L., Chen, Z., Shao, J.: Thinking in frequency: face forgery detection by mining frequency-aware clues. In: Vedaldi, A., Bischof, H., Brox, T., Frahm, J.-M. (eds.) ECCV 2020. LNCS, vol. 12357, pp. 86–103. Springer, Cham (2020). https://doi.org/10.1007/978-3-030-58610-2_6

21. Li, J.M., Xie, H.T., Li, J.H., et al.: Frequency-aware discriminative feature learning supervised by single-center loss for face forgery detection. In: Proceedings of 2021 IEEE/CVF Conference on Computer Vision and Pattern Recognition, pp. 6454–6463. IEEE, Nashville, USA (2021)

22. Wang, J.K., Wu, Z.X., Ouyang, W.H., et al.: M2TR: multi-modal multi-scale transformers for deepfake detection. In: Proceedings of the 2022 International Conference on Multi media Retrieval, pp. 615–623. ACM, Newark, USA (2022)

23. Conotter, V., Bodnari, E., Boato, G., Farid, H.: Physiologically based detection of computer generated faces in video. In: Proceedings of 2014 IEEE International Conference on Image Processing (ICIP), pp. 248–252. IEEE, Paris, France (2014). https://doi.org/10.1109/ICIP.2014.7025049

24. Li, Y., Chang, M.C., Lyu, S.: In ictu oculi: exposing ai created fake videos by detecting eye blinking. In: 2018 IEEE International Workshop on Information Forensics and Security (WIFS), pp. 1–7. IEEE (2018)

25. Hu, S., Li, Y., Lyu, S.: Exposing GAN-generated faces using inconsistent corneal specular highlights. In: ICASSP 2021–2021 IEEE International Conference on Acoustics, Speech and Signal Processing (ICASSP), pp. 2500–2504. IEEE (2021)

26. Amerini, I., Galteri, L., Caldelli, R., del Bimbo, A.: Deepfake video detection through optical flow based CNN. In: 2019 IEEE/CVF International Conference on Computer Vision Workshop (ICCVW), p. 12051207. IEEE, Seoul, Korea (South) (2020). https://doi.org/10.1109/ICCVW.2019.00152

27. Mittal, T., Bhattacharya, U., Chandra, R., Bera, A., Manocha, D.: Emotions don' t lie: an audio-visual DeepFake detection method using affective cues [EB/OL] (2020). https://arxiv.org/pdf/2003.06711.pdf. Accessed 08 May 2023

28. Deng, J., Guo, J., Zhou, Y., Yu, J., Kotsia, I., Zafeiriou, S.: Retinaface: single-stage dense face localisation in the wild. CoRR, abs/1905.00641 (2019)

29. Verkruysse, W., et al.: Remote plethysmographic imaging using ambient light. Opt. Express 16(26), 21434–21445 (2008)

30. Ciftci, U.A., Demir, I., Yin, L.J.: FakeCatcher: detection of synthetic portrait videos using biological signals. IEEE Trans. Pattern Anal. Mach. Intell. 3029287 (2020b). https://doi.org/10.1109/TPAMI.2020.3009287

31. Rössler, A., Cozzolino, D., Verdoliva, L., Riess, C., Thies, J., Niessner, M.: FaceForensics++: learning to detect manipulated facial images. In: Proceedings of the 2019 IEEE/ CVF International Conference on Computer Vision, pp. 1–11. IEEE, Seoul, Korea (South) (2019). https://doi.org/10.1109/ICCV.2019.00009

32. Dolhansky, B., et al.: The DeepFake detection challenge (DFDC) dataset [EB/OL] (2020). https://arxiv.org/pdf/2006.07397.pdf. Accessed 08 May 2023

33. Li, Y.Z., Yang, X., Sun, P., et al.: Celeb-DF (v2): a new dataset for deepfake forensics [EB/OL] (2023). https://arxiv.org/abs/1909.12962. Accessed 08 May 2023

34. Thies, J., Zollhofer, M., Stamminger, M., Theobalt, C., Nießner, M.: Face2Face: real-time face capture and reenactment of RGB videos. In: IEEE Conference on Computer Vision and Pattern Recognition, pp. 2387–2395, June 2016

35. Zhao, C., Lin, C., Chen, W., Li, Z.: A novel framework for remote photoplethysmography pulse extraction on compressed videos. In: The IEEE Conference on Computer Vision and Pattern Recognition (CVPR) Workshops, June 2018

36. Haliassos, A., Vougioukas, K., Petridis, S., et al.: Lips don't lie: a generalisable and robust approach to face forgery detection. arXiv:2012.07657v3 [cs.CV]. 15 August 2021

37. Li, J., Xie, H., Li, J., et al.: Frequency-aware discriminative feature learning supervised by single-center loss for face forgery detection. arXiv:2103.09096v1 [cs.CV], 16 March 2021

A Design of Hybrid Transactional and Analytical Processing Database for Energy Efficient Big Data Queries

Wenmin Lin[✉]

Alibaba Business School, Hangzhou Normal University, Hangzhou, China
wenmin.lin@hznu.edu.cn

Abstract. With the prominent development of cloud computing and pervasive computing, huge volume of big data is accumulated in an ever-increasing manner. To process such huge volume of big data in an energy efficient manner is a popular topic in both industry and academia area. In this work, we discuss how to implement a hybrid transactional and analytical processing database to provide energy efficient big data processing capability. More specifically, PostgreSQL (PG) database is an excellent solution for handling Online Transactional Processing (OLTP) workloads. For OLTP databases to process Online Analytical Processing (OLAP) queries, the traditional solution is to dump the data from PG to an OLAP database such as Greenplum for further analysis. Such solution faces the challenges of extra energy consumption, data island, data inconsistency, to name a few. Hybrid Transactional and Analytical Processing (HTAP) systems, on the other hand, support running both transactional and analytical processing workloads on the same database, which has been achieved great attention recently. In this work, we propose a design of HTAP database by enhancing the high available OLTP PG clusters to support OLAP workloads, via the massively parallel processing (MPP) architecture. In our MPP PG cluster, the data is not split and each PG server maintains an identical replica of the whole data. Moreover, to speed up the execution efficiency, we split the data into multiple virtual parts and each PG server within the cluster only scan the pre-assigned data partition. A set of experiments on the public TPC-H dataset are conducted to evaluate the feasibility of our proposal.

Keywords: Hybrid Transactional and Analytical Processing (HTAP) · Massive Parallel Processing (MPP) · PostgreSQL cluster

1 Introduction

Nowadays, with the fast development of the technology of Internet-of-Things, cloud computing, as well as pervasive computing, data is produced, collected and accumulated by end devices such as sensors at any time and at any where in an ever-increasing and exploded manner. The huge volume of big data resources have been widely adopted to build various cyper and physical system applications in various fields, such as military, healthcare, transportation, agriculture, to name a few [1,16–18].

In practice, each sensor node normally has limited capacity in terms of data storage and data processing. To store the big data as well as mine valuable knowledge from the huge volume of big data collected by distributed sensors, the data is typically transferred to a remote cloud center via a set of communication protocols for further analysis. In such scenario, database systems provide a set of powerful set of tools to handle the big data storage and big data analysis requirements with the big data resources.

Moreover, for big data storage and big data processing, different requirements are raised in terms of the backend technique support. On one hand, from the perspective of big data storage, the data collected by distributed sensors is normally structured. For example, for a smart shipping application, the data is normally consisting of several fixed dimension, such as <longitude, latitude, ship number, average speed, timestamp>. As a result, distributed relational database can be adopted to store such data for further analysis.

On the other hand, from the perspective of big data analysis, there are mainly two kinds of analytical workloads for big data collected by wireless sensor nodes: (1) "one-shot" analytical workloads which are frequently executed and normally involving limited volume of data items, (2) complex analytical workload which relates to huge volume of data items. In general, from the perspective of database, one-shot analytical workload belongs to typical Online Transactional Processing (OLTP) workload, for which the execution time is short and the workload is highly concurrent. While the complex analytical workload belongs to Online Analytical Processing (OLAP) workload, for which the execution frequency is low and involves multiple data items and takes much longer time compared with OLTP workload. Typically, there are two kinds of databases for handling OLTP workloads and OLAP workloads, i.e., OLTP database and OLAP database, respectively.

For applications that are required to handle both OLTP and OLAP workloads, the typical solution is to maintain an OLTP and OLAP database separately. The OLTP database handles real-time and one-shot OLTP workloads. And for complex big data analysis, the data will be transferred from OLTP database to OLAP database for further analysis via a set of ETL tools. Such solution faces limitations of extra energy consumption, data inconsistency, lack of real-time response, to name a few. Therefore, to store and process the big data in an energy efficient manner, we could try to explore new solutions to support handing both OLTP and OLAP queries in a more energy efficient manner.

In view of these observations, exploring HTAP database to support both OLTP and OLAP workload on a singe host has become a hot research topic among both industry and academia area. Currently, there have been quite a few practices on HTAP database [10] to support both OLTP and OLAP queries on a single cluster. Among them, the typical solution is Greenplum. For example. Lyu [3] et al. try to enhance the native Greenplum database with OLTP capabilities with a deadlock detector and one-phase commit protocol. However, Greenplum is designed to handle OLAP workloads in native and it adopts the shared-nothing structure to handle big data queries. And adopting such structure to support OLTP queries will suffer from scalability issue.

In this paper, we are aiming at designing a HTAP database to support the big data analysis in terms of both OLTP queries and OLAP queries on the same database system. More specifically, the HTAP database will support both Online Transaction Processing (OLTP) queries as well as Online Analytical Processing (OLAP) workloads on the identical single PG database cluster. The PG database cluster to support OLTP queries in native is adopted, where the extra resources on the PG database cluster is taken advantage to handle OLAP workloads.

The contribution of this work is summarized as follows:

- We propose a design of HTAP database based on the OLTP PG clusters, where the idle standby PG servers are reused to handle OLAP workloads;
- The Massively Parallel Processing architecture is adopted, where data is virtually split and assigned to corresponding individual PG sever, to improve the query efficiency;
- The Motion operator in PG is customized to hide the virtual data partition, and shuffle data between PG servers when necessary.

The reminder of this paper is organized as follows. Section 2 discusses the related work on HTAP databases and the MPP architecture. Section 3 introduces our work on the HTAP database based on the native PG cluster. In Sect. 4, we evaluate the performance of our proposal by comparing with the native PG server. Section 5 concludes the paper and points out the future work.

2   Related Work

We are discussing how to implement a database system to support HTAP queries (i.e., both OLTP queries and OLAP quires) on an identical database cluster in this work. The HTAP database system takes advantage of the idle resources on native PG database cluster via the MPP architecture. Therefore, we are going to review existing works on the HTAP databases, the Massively Parallel Processing architecture, respectively.

2.1   The HTAP Databases

Hybrid Transactional and Analytical Processing (HTAP) systems have gained significant attention in the past decade. HTAP systems combine both Online Transactional Processing (OLTP) and Online Analytical Processing (OLAP) systems into a single, unified system, which can provide real-time analytical query processing over fresh, transactional data. Due to its benefits that making real-time data analysis available without extra components or external systems, as well the reduction of overall business cost in terms of hardware and administration, HTAP systems have become popular in the past decade and have gained wide attention from both industry and academia.

There is a significant amount of existing works on HTAP systems and architectures. From the perspective of HTAP architectures, existing HTAP architectures could be divided into three categories: [5]: single-copy, mixed-format (SCMF), two-copy, mixed-format (TCMF), and single-copy, single-format (SCSF).

In the SCMF architecture, a single copy of main data is kept by the database, and an intermediate data structure delta is used for OLTP storage. The main data and the delta data is in different data format. Moreover, the delta is only modified by OLTP transactions; while the OLAP queries can read both the main copy of the stored data as well as the delta. The delta is maintained in row format, whereas the main copy of the data is in columnar format. Examples of SCMF HTAP systems include SAP HANA [6] and MemSQL implementation [7].

In the TCMF architecture, two copies of the data is kept, among which one is in row format for the OLTP queries, and another one is in columnar format for OLAP queries. Moreover, to keep consistency for the data copies across the OLTP and OLAP components, similar to the SCMF architecture, TCMF also adopts a delta data structure, which keeps track of the recently modified data items. The recently updated fresh tuples are periodically propagated from the OLTP side to the OLAP side by scanning the delta data structure. In practice, typical example of TCMF HTAP systems is BatchDB [11].

While the SCSF architecture keeps only a single copy of the data with a single data format (i.e., only row or columnar) for both OLTP and OLAP workloads. In SCSF [12,13], copy-on-write snapshooting and multi-version concurrency control mechanism are adopted to keep multiple versions of the data. Moreover, with such SCSF architecture, the OLAP queries can always access the most recent transactional data.

Although there have been quite a few HTAP databases have been studied and deployed, but the combination of HTAP database and the big data analysis requirement over WSNs still lacks extensive research. In this paper, we focus on designing a cloud-based HTAP database to support hybrid big data analysis queries over the wireless sensor networks. Compared with existing practices of HTAP databases, our HTAP database is based on a cluster of high concurrent and high available PostgreSQL (PG) instances with SCSF architecture, where only one data copy is maintained with a single data format. The WAL log copy is adopted to keep data in consistency among all the PG instances. The design goal of our work is to improve the resource utilization rate and scalability of the HTAP databases.

2.2   Massively Parallel Processing Architecture

The *Massively Parallel Processing* (MPP) architecture is widely adopted in multiple database systems to provided parallel data query functionality. Typical MPP databases include Greenplum, GBase, Netezza, and Teradata [4], to name a few. Those MPP databases are designed for handling query processing over distributed relational big data in native. They distribute pipelined query processing across multiple worker segments. And the data storage of the multiple segments adopts a shared-nothing architecture. In the shared-nothing architecture, a single coordinator node handles communication with clients and query optimization. The MPP systems are normally achieving high-performance, but are not scale well for large clusters.

In a MPP architecture based database cluster, there is only one host called the coordinator host in the whole database system. The others are called worker hosts for short. The coordinator host directly connects to the user clients to receive commands or queries. Moreover, it will generate a distributed query plan by the query optimizer, dispatches the plan to each worker host, gather the results, and finally sends back to the clients. The worker instances on the coordinator host storage some meta data. Moreover, in MPP architecture, the data storage follows a shared-nothing architecture. The coordinator host and worker host have their own shared memory and data directory. The data is partitioned into multiple pieces and each worker host maintains only one piece of the whole data. In other words, the worker hosts serve as the primary storage of user data and execute a specific part of the distributed plan that is dispatched from the coordinator host.

Similar to existing MPP databases, we will adopt the MPP architecture to build the HTAP database in this work. Different from the existing MPP framework with shared-nothing structure, in our proposal, all the standby PG nodes shares the same data across the distributed cluster. Compared with the shared-nothing architecture, such design could achieve better scalability, and can be easily migrated to applications with shared disk storage model.

3   The Framework of MPP Database Based on Shared PG Clusters

As shown in Fig. 1, the whole architecture of the HTAP database is implemented by reuse the idle standby PG servers in the traditional high available OLTP PG cluster.

The MPP cluster in our proposal consists of a set of standby PG servers across multiple hosts, where one node will be picked up as the coordinator and the rest standby PG servers are worker segments. The users connect to the coordinator node to execute its queries; while the back-end distributed PG servers are transparent to the users. On receiving user queries, the coordinator will generate a distributed query plan, and dispatches the query plan to corresponding PG instances. After each instance finishes the plan, it will send back the local execution results to the coordinator node to generate the final result, which will be finally sent back the users. The typical Volcano execution model is adopted here, as shown in Fig. 2.

We take a shared-storage architecture in our proposal. Different from the Greenplum design which adopts a shared-nothing architecture, the coordinator and PG instances across the cluster share their storage. Moreover, to guarantee

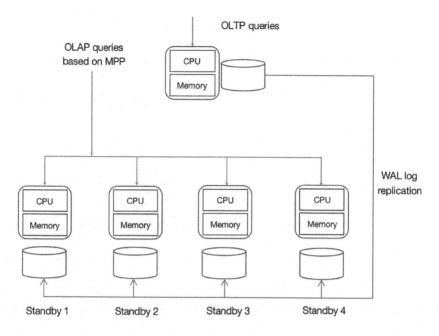

Fig. 1. The architecture to handling OLAP workloads

the consistency regarding data storage across multiple servers, the coordinator is responsible for replicating the WAL logs across the cluster. Each PG instance will replay the logs to guarantee the consistency of its local data storage.

Since the data on each segment server is identical with each other, the OLAP workloads can be theatrically assigned to a single host for execution. However, such solution will not take full advantage of the multiple PG servers within the cluster. For each user query, the distributed planner will generate a distributed query plan. Different from the native planner adopted in PG, we adopt the open-source ORCA optimizer to generate the distributed query plan. Moreover, since the PG cluster has a shared data storage architecture, we will virtually split the relational dataset into multiple disjoint partitions, and each PG segment will only scan a pre-assigned virtual dataset.

3.1   The OLTP Workload Manager

In our proposal, each server within the cluster is an PG instance, and the transaction can be committed or abort on the PG servers concurrently. Since PG supports OLTP workload in native, we will not unfold the details in this work. However, the multiple servers should achieve consistency regarding the under layer data storage. For this problem, the consensus protocol such as Paxos, Raft would be adopted. Since our major work is to add OLAP features into the native OLTP PG database, we will leave this issue to be discussed in future work.

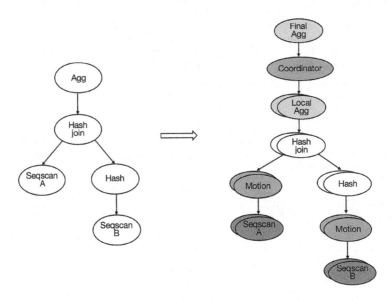

Fig. 2. The query plan on a single host vs. the distributed query across the PG cluster

3.2   The Distributed Planner and Distributed Executor

Since the data is virtually partitioned and each segment will only scan the pre-assigned data partition, we could view the MPP PG cluster as a Greenplum-like database. For a given OLAP query that joins two relations for example, we need to check whether the two tuples from different PG servers match the join condition. As a result, the data partition could be moved across multiple PG servers to obtain all possible matching tuples. Similar to the Motion operator introduced in Greenplum, we optimized the Motion operator in our design to hide the under laying virtual partition and move the data when necessary.

An example of the MPP architecture adopted in this paper is shown in Fig. 3. In Fig. 3, there are three PG severs: one is the coordinator node, and the remaining two servers are worker segments. For a given user query containing a join SQL, a distributed plan is generated and dispatched to each node. The Motion operator naturally cut the plan into multiple pieces, each piece below or above the Motion operator is called a slice. In Fig. 3, the top slice is executed by the coordinator node; while the other slices are executed by the PG segments. Each PG segment will scan its local virtual data partition and sends the tuples out using the Motion operator. On receiving the tuples, a segment that performs hash join will scan its local table to build a hash table to produce the hash join results, so as to generate the local aggregation results. Finally, the local result will be sent to the coordinator node to generate the final result.

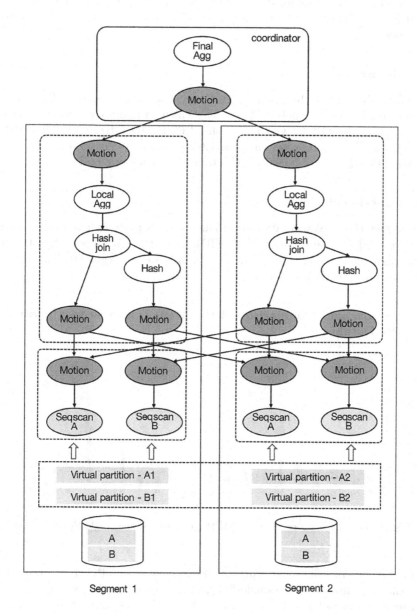

Fig. 3. The architecture to handling OLAP workloads

4   Evaluation

In this section, we evaluate the performance of the proposed HTAP database based on a PG cluster via a set of experiments to evaluate the effectiveness of our proposal.

4.1   Dataset

We adopt the TPC-H benchmark [19] to evaluate our proposal. The TPC-H queries consists of 22 analytical queries, which are running against datasets of two scales: 160 GB (10 GB/node) and 1.6 TB (100 GB/node). The 160 GB dataset is a CPU-bound case while the 1.6 TB case is IO-bound. In our case, we run the TPC-H queries on the 160 GB to evaluate our proposal.

4.2   Experimental Setting

We conduct the implementation on a cluster of 16 hosts (where each host has 256 cores and storage capacity of 128 GB), among which 1 node plays the role of coordinator; and remaining 15 nodes works as worker segments. The network bandwidth is 400 MB/s.

4.3   Performance Comparison of the MPP Database vs. the Single PG Database

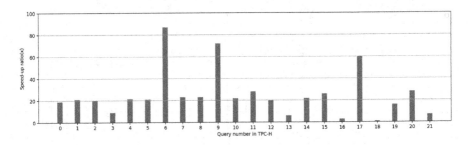

Fig. 4. The speed-up ratio of the MPP database vs. the single PG database

In this experiment, we execute the 22 SQL queries TPC-H benchmark on both the PG cluster and the single PG host, and compare the execution latency between the two solutions. As shown in Fig. 4, for the 22 SQL queries, most of them are achieving better performance when executing on the PG clusters. Moreover, there are 3 SQL queries achieving more than 60 times speed-up ratio. Also, there are about 19 queries are achieving more than 10 times speed-up ratio. The experimental results prove that the MPP database could achieve better OLAP performance by comparing with the native PG cluster.

4.4   Performance Comparison w.r.t the Number of CPU Cores

Moreover, we test the system performance of the MPP PG cluster with respect to different number of CPU cores. All the 22 queries are executed and average latency is recorded with different experimental setting. As shown in Fig. 5, when the number of CPU cores are less than 128 cores, with the increase of the number of CPU cores, the average execution latency is reduced linearly. And when the number of CPU cores reaches 256, the latency tends to be stable around 15 min. Therefore, such MPP design could sufficiently take advantage of the CPU resources on the PG cluster.

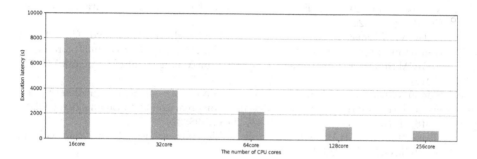

Fig. 5. The execution latency of the MPP database w.r.t different number of CPU cores

5   Conclusion

In this paper, we proposes a design of HTAP database based on a cluster of native PG servers via the massively parallel processing (MPP) architecture. In our MPP PG cluster, the data is not split and each PG server maintains an identical replica of the whole data. Moreover, to speed up the execution efficiency, we split the data into multiple virtual parts and each PG segment only scan the pre-assigned data part. The experimental results show that our proposal could achieve better performance than the native PG servers. Moreover, it could achieve better performance with the increase of the number of CPU cores. Due to the limitation of paper length, the details of how to achieve consistency regarding the data storage across multiple servers are not discussed in this work. Moreover, the comparison between our proposal and existing OLAP databases such as Greenplum will be analysed in our future work as well.

Acknowledgement. This work was supported in part by the Natural Science Foundation of Zhejiang Province (No. LQ21F020021) and Research Start-up Project funded by Hangzhou Normal University (No. 2020QD2035).

References

1. Xu, X., Fang, Z., Zhang, J., et al.: Edge content caching with deep spatiotemporal residual network for IoV in smart city. ACM Trans. Sens. Netw. **17**(3), 1–33 (2021)
2. Graefe, G.: Volcano - an extensible and parallel query evaluation system. IEEE Trans. Knowl. Data Eng. **6**(1), 120–135 (1994)
3. Lyu, Z., Zhang, H.H., Xiong, G., et al.: Greenplum: a hybrid database for transactional and analytical workloads. arXiv arXiv:2103.11080 (2021)
4. Arnold, J., Glavic, B., Raicu, I.: A high-performance distributed relational database system for scalable OLAP processing. In: 2019 IEEE International Parallel and Distributed Processing Symposium (IPDPS), pp. 738–748 (2019)
5. Sirin, U., Dwarkadas, S., Ailamaki, A.: Performance characterization of HTAP workloads. In: 2021 IEEE 37th International Conference on Data Engineering (ICDE), pp. 1829–1834 (2021)
6. Sikka, V., Färber, F., Lehner, W., Cha, S.K., Peh, T., Bornhövd, C.: Efficient transaction processing in SAP HANA database: the end of a column store myth. In: ACM SIGMOD International Conference on Management of Data, pp. 731–742 (2012)
7. Skidanov, A., Papito, A.J., Prout, A.: A column store engine for real-time streaming analytics. In: IEEE 32nd International Conference on Data Engineering (ICDE) (2016). https://doi.org/10.1109/ICDE.2016.7498332
8. Chang, L., et al.: HAWQ: a massively parallel processing SQL engine in hadoop. In: SIGMOD, pp. 1223–1234 (2014)
9. Thusoo, A., et al.: Hive-aware housing solution over a map-reduce framework. In: Proceedings of the VLDB Endowment, pp. 1626–1629 (2009)
10. Sirin, U., Dwarkadas, S., Ailamaki, A.: Performance characterization of HTAP workloads. In: 2021 IEEE 37th International Conference on Data Engineering (ICDE), pp. 1829–1834 (2021)
11. Makreshanski, D., Giceva, J., Barthels, C., Alonso, G.: BatchDB: efficient isolated execution of hybrid OLTP+OLAP workloads for interactive applications. In: 2017 IEEE 33nd International Conference on Data Engineering (ICDE), pp. 37–50. 585 (2017)
12. Appuswamy, R., Karpathiotakis, M., Porobic, D., Ailamaki, A.: The case for heterogeneous HTAP. In: CIDR, pp. 1041–1052 (2017)
13. Raza, A., Chrysogelos, P., Anadiotis, A.C., Ailamaki, A.: Adaptive HTAP through elastic resource scheduling. In: 2020 ACM International Conference on Management of Data, pp. 2043–2054 (2020)
14. Oracle Exadata. https://www.oracle.com/technetwork/database/exadata/exadata-storage-technical-overview-128045.pdf
15. Verbitski, A., Gupta, A., et al.: Amazon Aurora: design considerations for high throughput cloud-native relational databases. In: Proceedings of the 2017 ACM International Conference on Management of Data, pp. 1041–1052 (2017)
16. Qi, L., Dou, W., Chunhua, H., Zhou, Y., Jiguo, Yu.: A context-aware service evaluation approach over Big Data for cloud applications. IEEE Trans. Cloud Comput. **8**(2), 338–348 (2020)
17. Xu, X., et al.: A computation offloading method over Big Data for IoT-enabled cloud-edge computing. Futur. Gener. Comput. Syst. **95**, 522–533 (2019)
18. Zhang, X., et al.: MRMondrian: scalable multidimensional anonymisation for Big Data privacy preservation. IEEE Trans. Big Data (2017). https://doi.org/10.1109/TBDATA.2017.2787661
19. TPC-H. http://www.tpc.org/tpch/

Chinese Medical Named Entity Recognition Based on Pre-training Model

Fang Dong[1], Shaowu Yang[1(✉)], Cheng Zeng[2], Yong Zhang[3], and Dianxi Shi[4]

[1] National University of Defense Technology, Changsha 410073, China
fangfangdong2000@163.com
[2] Wuhan University, Wuhan 430072, China
[3] SD steel Rizhao Co., Ltd., Shandong, China
[4] Tianjin Artificial Intelligence Innovation Center, Tianjin 300457, China

Abstract. Named Entity Recognition (NER) task aims to identify named entities from unstructured text and classify them into corresponding entity types. Existing pretraining models typically utilize BERT models to learn word embeddings at the character level, disregarding the semantic relationships between phrases. They also pay less attention to long-distance dependencies within sentences. Additionally, the datasets suffer from challenges such as small scale, lack of standardization, and annotation errors, all of which contribute to poor model robustness. Therefore, this paper proposes a Chinese medical named entity recognition model based on RoBERTa (A Robustly Optimized BERT Pre-training Approach), adversarial training, and hybrid encoding layers to enhance semantic understanding and model generalization. The proposed model is evaluated on three real clinical datasets. And experimental results demonstrate significant performance improvement compared to the baseline models. Furthermore, the advantages of this proposed approach over the baseline models are further analyzed through experiments.

Keywords: Chinese Medical Named Entity Recognition · Pre training model · BiLSTM · R-Transformer · Natural Language Processing

1 Introduction

Named Entity Recognition (NER) is a prominent research branch in natural language processing, which was introduced at MUC-6 (The Sixth Message Understanding Conference) in 1996 [1]. Biomedical Named Entity Recognition (BioNER) is a subfield of NER.

Compared to NER tasks in general domains, biomedical named entities (BioNEs) often exhibit longer and more complex vocabulary. Moreover, there

This work was supported by the Integrated Program of National Natural Science Foundation of China (No.91948303).

are often multiple variants or abbreviated names for a single entity, and the same entity may represent different meanings. Therefore, biomedical NER tasks are more challenging than NER tasks in general domains.

For Chinese corpora, Chinese is more complex than English in terms of character composition and sentence structure. These further increases the difficulty of entity recognition. For example, "复杂性肝癌切除+胆囊切除+胆总管切开取癌栓+胆道镜探查+T管引流术". Additionally, Chinese texts lack natural word segmentation similar to English, so preprocessing operations like word segmentation are required before entity recognition, introducing additional challenges to the task.

Existing methods for Chinese biomedical NER can be classified into three main categories: 1) dictionary-based and rule-based methods that rely on manually crafted rules; 2) statistics-based machine learning methods, which learns annotation models from large-scale corpora; 3) deep learning methods, which have been the focus of recent research. These methods combine pre-training models like BERT with neural network models.

This paper adopts a deep learning approach to complete entity extraction tasks, thereby improving annotation efficiency, accuracy, and significantly reducing manual costs. The main contributions of this paper are as follows:

1. Data integration: This paper compiles annotated datasets from the Medical Named Entity Recognition tasks in CCKS (China Conference on Knowledge Graph and Semantic Computing) from 2017 to 2020 and at the 2020 CHIP (China Health Information Processing Conference). The data annotations were unified and standardized, ensuring the effectiveness of the integrated dataset.
2. Model construction: The paper proposes a Chinese biomedical named entity recognition model based on a combination of RoBERTa pre-training model and BiLSTM with R-Transformer as a hybrid encoding layer. Additionally, adversarial samples are introduced during training, to enhance the model's robustness. On the constructed dataset, the proposed model achieves a 1.43%-2% improvement in F1 scores compared to other baseline methods, demonstrating good performance. Furthermore, ablative experiments were conducted, further confirming the effectiveness of the proposed model.

The remaining parts of this paper are organized as follows: Sect. 2 introduces related works, Sect. 3 describes the Chinese biomedical named entity recognition model based on the pre-training model, Sect. 4 presents experimental details and discusses the results, and Sect. 5 concludes the paper.

2 Related Work

2.1 Named Entity Recognition

Common Chinese medical named entity recognition (NER) methods can be divided into two categories: statistical machine learning methods and deep learning methods.

Statistical machine learning methods include widely used approaches such as Support Vector Machines (SVM) [2,3], Hidden Markov Models (HMM) [4], Conditional Random Fields (CRF) [5], and Maximum Entropy Models (ME) [6]. Ye, Lei [7,8] constructed some labeled datasets of Chinese electronic medical records and conducted research on them. However, traditional machine learning methods require manual feature selection and often yield suboptimal recognition results.

In recent years, deep learning-based named entity recognition methods have been widely adopted. Compared to traditional machine learning models, deep neural networks require less manual feature engineering and achieve significant improvements in accuracy and recall. Common network architectures include Convolutional Neural Networks (CNN) [9,10], Recurrent Neural Networks (RNN) [11], Long Short-Term Memory Networks (LSTM) [12], using CRF as an output layer. Additionally, methods like word embeddings, attention mechanisms, and other techniques are employed to further improve the algorithm's precision. Wang et al. [13] incorporated dictionaries into deep neural networks for Chinese NER. Xu et al. [14] proposed a medical NER model based on Bidirectional LSTM [16] and CRF. Tang et al. [15] introduced an attention-based CNN-LSTM-CRF model for entity recognition in Chinese clinical texts. CCKS held information extraction competitions from 2017 to 2020, which included medical named entity recognition tasks. Most participants used combined approaches of pre-trained models and neural networks, such as BiLSTM-CRF or BERT-CRF [16].

2.2   Pre-trained Models

Pre-trained models refer to training models using a large amount of unrelated data in advance, which are independent of subsequent tasks. In the Chinese medical named entity recognition task, researchers have gradually focused on pre-trained models due to the lack of a significant amount of medical text data with entity annotations. Pre-trained models can acquire prior semantic knowledge from a large amount of unlabeled text, thereby alleviating the training burden downstream.

In the early days, word embedding training commonly used models such as word2vec [17] and GloVe [18]. Ramachandran et al. [19] proposing the Sequence-to-Sequence (Seq2Seq) approach. However, these methods utilize single-layer neural networks and are unable to understand word meanings in different contexts [20]. In 2018, Peters et al. [21] introduced the ELMo model. Subsequently, pre-trained models such as GPT [22] and BERT [16] were proposed, with BERT achieving state-of-the-art performance in various NLP tasks. However, the BERT model still has certain limitations that need improvement. Liu et al. [23] improved the BERT model and proposed the RoBERTa model, enhancing the model's ability to recognize vocabulary at the phrase level by modifying the masking method.

3   Chinese Medical Named Entity Recognition Based on Pre-training Model

The Chinese medical named entity recognition model based on pre-trained models consists of three main parts: the embedding layer, the encoding layer, and the normalization layer. The embedding layer loads the pre-trained model parameters and converts the input text sequence into a matrix of vectors. The encoding layer utilizes a hybrid network model combining BiLSTM and R-Transformer [24] to learn the encoding of word embeddings and concatenates the results to obtain probability predictions for each label classification. The normalization layer uses the CRF algorithm to correct the output of the upper layers and obtain the final label sequence. The model structure is shown in Fig. 1.

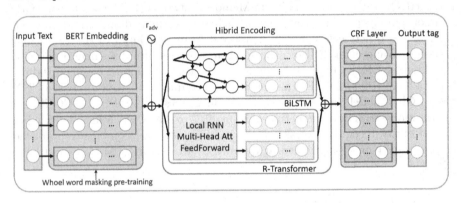

Fig. 1. Structure diagram of Chinese medical Named Entity Recognition based on pre training model.

Chinese medical named entity recognition is also a sequence labeling task with BIO (Begin, Inside, Outside) tags. The model uses RoBERTa, a pre-trained model trained on a large amount of unrelated Chinese data, to generate word embeddings. The word vector representations, along with perturbed samples generated based on those representations, are used as inputs to the encoding layer. They are separately passed through BiLSTM and R-Transformer for feature extraction of word vectors. Finally, the outputs from the hidden layers are combined, and the CRF is applied to normalize the final annotation results, avoiding logical misclassifications and obtaining the ultimate sequence labeling results.

3.1   Pre-training

In this paper, the RoBERTa-wwm-ext-large pre-trained model is used to convert input text into word vector representations. RoBERTa is a variant of the BERT model and inherits its structure, consisting of 24 layers of Transformer with a multi-head self-attention head size of 16 and 1024 hidden units per layer.

RoBERTa improves upon the masking mechanism of the BERT model, allowing the improved model to learn more semantic information compared to BERT. Additionally, this pre-trained model utilizes a larger Chinese corpus for pre-training, further enhancing the expressive power of the model for word embeddings.

During training, BERT divides the input sequence into subword units at the character level for Chinese. It replaces 15% of the words in the sentence with the [Mask] token for prediction. However, for Chinese, a complete expression of the smallest semantic unit often requires two or more characters. This training approach often masks a character within a word or phrase, resulting in a loss of internal semantic, grammatical, and syntactic information within the sentence. RoBERTa integrates phrase-level knowledge into language expression during training, combining prior knowledge with vocabulary and semantic information, leading to better performance in knowledge-driven tasks. The masking strategy is shown in Fig. 2.

Sentence	Harry	Potter	is	a	series	of	fantasy	novels	written	by	British	author	J.	K.	Rowling
Basic-level Masking	[mask]	Potter	is	a	series	[mask]	fantasy	novels	[mask]	by	British	author	J.	[mask]	Rowling
Pharse-level Masking	Harry	Potter	is	a	series	[mask]	fantasy	novels	[mask]	by	British	author	[mask]	[mask]	[mask]

Fig. 2. Different levels of concealment in sentences.

Firstly, RoBERTa retains the basic-level masking mechanism used in the BERT model. It treats a sentence as a basic unit and masks 15% of the language units at the character level. The stacked Transformer encoders are trained to predict the masked tokens, allowing the model to learn basic word representations. In addition, RoBERTa introduces a phrase-level masking mechanism. A phrase is a conceptual unit composed of a group of English words or Chinese characters. English phrases are obtained through lexical analysis and chunking tools, while Chinese phrases are obtained through relevant word segmentation techniques. During training, a sentence is still treated as a basic language unit sequence, and random phrases within the sentence are masked and predicted. This allows the model to encode semantic information between phrases into word embeddings. The matrix W obtained from training is used as the initial lookup table for word embeddings in this paper.

3.2   Data Collection and Processing

In this paper, we collected datasets from the CCKS2017-2020 competitions on Chinese medical named entity recognition tasks, as well as the CHIP2020 competition dataset. Firstly, the differences in named entity recognition among different datasets were standardized, and samples with erroneous text content or annotations were removed. Next, based on the provided information of medical named entities present in the datasets, the boundaries of some medical named entities in the datasets were revised. After completing these preparations, the original text and sequence annotation were aligned using the BIO tagging format, preparing the data for model training.

3.3 Embedding Layer

RoBERTa (A Robustly Optimized BERT Pretraining Approach) is a pre-trained model based on BERT that incorporates improvements. RoBERTa removes the next sentence prediction task from BERT and focuses more on semantic connections within a single text. It employs a dynamic masking mechanism to enable the model to learn different language representations and optimizes the details of the BERT model parameters. The parameters of RoBERTa can be directly loaded using the BERT model, and the model architecture is shown in Fig. 3.

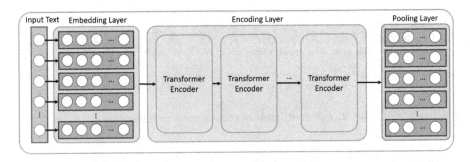

Fig. 3. Model Architecture of RoBERTa.

Let the input text be denoted as $s = \{s_1, s_2, \ldots, s_n\}$, where s_i represents the i-th character in the sentence. During input processing, the starting position of each sentence is marked with the $[CLS]$ token, and the end of each sentence is marked with the $[SEP]$ token, forming the input sequence $S = \{s_0, s_1, s_2, \ldots, s_n, s_{n+1}\}$. Here, $s_0 = [CLS]$ and $s_{n+1} = [SEP]$. The input sequence S goes through three types of embeddings in the embedding layer: Token, Segment, and Position embeddings, as shown in formulas (1)–(3).

$$Token_i = TOK(s_i) \tag{1}$$

$$Seg_i = SEG(TOK(s_i)) \tag{2}$$

$$Pos_i = POS(i) \tag{3}$$

Here, $TOK(\cdot)$ maps the input to the word embedding matrix W, which is obtained during the pretraining process. It retrieves the word embedding corresponding to each input character from the matrix W, resulting in the Token matrix $T = \{Token_0, Token_1, \ldots, Token_{n+1}\}$. $SEG(\cdot)$ is used to indicate the sentence position of the current input sequence in the text. $POS(\cdot)$ assigns unique positional encoding to each vector based on the order of the input sequence. Which aims to ensure that the initial position embeddings of different positions within the same sequence are distinct and the initial position embeddings of the same positions across different sequences are the same. Finally, based on these three results, the output of the embedding layer $E = \{E_0, E_1, \ldots, E_{n+1}\}$ is calculated as shown in formula (4).

$$E_i = Token_i + Seg_i + Pos_i \tag{4}$$

Next, the output E of the embedding layer undergoes computations in a multi-layer Transformer encoder to obtain the new output $Trans = Trans_0, Trans_1, \ldots, Trans_{n+1}$, as shown in formulas (5) and (6).

$$temp_i = Norm(Dropout(MultAttention(E_i)) + E_i)V \tag{5}$$

$$Trans_i = Norm(Dropout(FeedForward(temp_i)) + temp_i) \tag{6}$$

Finally, the final output matrix $Trans'$ is obtained by applying an activation function through the pooling layer.

The Token matrix T, the output E of the embedding layer, the output matrix $Trans$ obtained through the Transformer calculations, and the final word embedding matrix $Trans'$ are then fused using a weighted blending approach, resulting in the final word embedding matrix of the input text, as shown in formula (7).

$$Outputs_i = Trans'_i \times weight \tag{7}$$

In the formula, $weight$ represents a pre-defined weight matrix.

3.4 Hybrid Encoding Layer

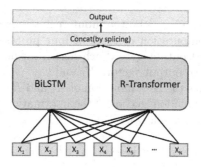

Fig. 4. Model Architecture of Hybrid Encoding Layer.

This model combines two encoders, namely Bidirectional Long Short-Term Memory (BiLSTM) and R-Transformer, as the core structures of the encoding layer. The two models receive the word embedding matrix $C = \{C_0, C_1, \ldots, C_{N+1}\}$ as input and generate output matrices $B = \{B_0, B_1, \ldots, B_{N+1}\}$ for BiLSTM and $R = \{R_0, R_1, \ldots, R_{N+1}\}$ for R-Transformer through encoding operations. Finally, the two output matrices are concatenated to obtain the final output matrix of the encoding layer. The model structure of the hybrid encoding layer is shown in Fig. 4.

BiLSTM. The introduction of Long Short-Term Memory (LSTM) greatly alleviated the issues of gradient vanishing and gradient exploding in RNNs. BiLSTM, as a variant of LSTM, learns from sequences in both forward and backward directions, allowing it to capture the contextual information of the text. It consists of two independent LSTM networks. The input sequence is fed into the two LSTM models separately in the forward and backward directions. The outputs of the two LSTM models, h_t, are added, averaged, or concatenated to obtain the final output of the BiLSTM model. The individual core structure is shown in Fig. 5.

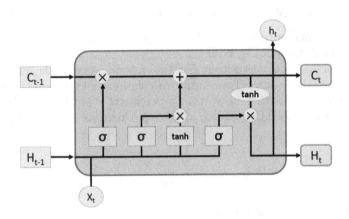

Fig. 5. Architecture of LSTM Core.

Let $C = \{C_0, C_1, \ldots, C_{N+1}\}$ be the input vector representation of BiLSTM. Each LSTM kernel computation consists of four processes: forget gate, input gate, state update, and output gate. Each kernel includes three input vectors X_t, H_{t-1}, C_{t-1}, and two output vectors H_t and C_t.

The forget gate takes the concatenation of X_t and H_{t-1} as input and applies the sigmoid activation function to control the information retention of C_{t-1}. The calculation is shown in formula (8).

$$f_t = \sigma(W_f \cdot [h_{t-1}, x_t] + b_f) \tag{8}$$

The input gate takes the concatenation of X_t and H_{t-1} as input and performs calculations according to formulas (9) and (10).

$$i_t = \sigma(W_i \cdot [h_{t-1}, x_t] + b_i) \tag{9}$$

$$\widetilde{C_t} = tanh(W_c \cdot [h_{t-1}, x_t] + b_c) \tag{10}$$

The state update combines C_{t-1} multiplied by the calculation result of the forget gate, f_t, and the new candidate value $i_t \times \widetilde{C_t}$ generated by the input gate to obtain the new kernel state C_t. The calculation is shown in formula (11).

$$C_t = f_t \times C_{t-1} + i_t \times \tilde{C}_t \tag{11}$$

The output gate takes the concatenation of X_t and H_{t-1}, and the kernel state C_t as input and produces the final output h_t through function computations. The calculation is shown in formulas (12) and (13).

$$o_t = \sigma(W_o \cdot [h_{t-1}, x_t] + b_o) \tag{12}$$

$$h_t = o_t * tanh(C_t) \tag{13}$$

R-Transformer. The R-Transformer network structure consists of three parts: LocalRNN, Multi-Head Self-Attention, and Feed-Forward Neural Network.

LocalRNN is a self-adjusting recurrent neural network sequence that processes input sequences of the same length as its depth. It processes the sequences in the same way for each consecutive subsequence, producing the final hidden state matrix. This approach enhances the capture of local and sequential information in the language sequence. And in sliding window mode, the global sequential information is not lost, effectively mitigating the gradient-related issues of original RNNs. The calculation is shown in formulas (14) and (15).

$$h_t = LocalRNN(x_{t-M-1}, x_{t-M-2}, \ldots, x_t) \tag{14}$$

$$h_1, h_2, \ldots, h_N = LocalRNN(x_1, x_2, \ldots, x_N) \tag{15}$$

Here, M represents the predetermined length of the subsequence.

The Multi-Head Self-Attention layer takes the output matrix of the Local-RNN layer, added to the original input matrix X and normalized, as input and performs multi-head self-attention calculations. The calculation is shown in formulas (16) and (17).

$$u_t = MultiHeadAttention(h_1, h_2, \ldots, h_t) \tag{16}$$

$$= Concatenation(head_1(h_t), head_2(h_t), \ldots, head_k(h_t))W^0 \tag{17}$$

Where $head_k(h_t)$ is the result of the k-th self-attention calculation, W^0 is a linear projection matrix. Specifically, $head_k(h_t)$ is the weighted sum of all value vectors, where the weights are calculated by applying attention functions to all query-key pairs, as shown in Eqs. (18) and (19).

$$\alpha_1, \alpha_2, \ldots, \alpha_n = softmax(\frac{qk_1^T}{\sqrt{d_k}}, \frac{qk_2^T}{\sqrt{d_k}}, \ldots, \frac{qk_n^T}{\sqrt{d_k}}) \tag{18}$$

$$head_i(h_t) = \sum_{j=1}^{n} \alpha_j v_j \tag{19}$$

Here, q, k_i and v_i are the query, key, and value vectors, respectively, and d_k is the dimension of vector k_i. The vectors k_i and v_i are obtained by projecting the input vector onto the query, key, and value matrices, as shown in Equation(20).

$$q, k_i, v_i = W^q h_t, W^k h_i, W^v h_i \tag{20}$$

W^q, W^k and W^v are projection matrices.

Finally, the feed-forward neural network layer takes the output matrices of the multi-head self-attention layer and the output matrix of the LocalRNN layer, adds them, normalizes the result. Then applies a non-linear transformation to the input matrix, as shown in Equation(21).

$$FeedForward = relu(xW_1 + b_1)W_2 + b_2 \tag{21}$$

Where x is the input matrix, W_1 and W_2 are used for expansion and contraction of the input matrix, b_1 and b_2 represent biases.

3.5 Normalization Layer

Conditional Random Field (CRF) is a discriminative probabilistic model that represents a Markov random field of output variables given a set of input variables. CRF combines the characteristics of maximum entropy models and hidden Markov models and is widely used in recognition tasks to constrain the final output.

Assuming a linear chain conditional random field $P(X \mid Y)$, the input variables X are $\{X_1, X_2, \ldots, X_n\}$, and the output sequence Y is $\{Y_1, Y_2, \ldots, Y_n\}$. The conditional probability of random variable Y taking the value$y$ given the input variable X taking the value x is as shown in Equations(22)and(23).

$$P(Y|X) = \frac{1}{Z(x)} exp(\sum_{i,k} \lambda_k t_k(y_{i-1}, y_i, x, i) + \sum_{i,l} \mu_l s_l(y_i, x, i)) \tag{22}$$

$$Z(x) = \sum_{y} exp(\sum_{i,k} \lambda_k t_k(y_{i-1}, y_i, x, i) + \sum_{i,l} \mu_l s_l(y_i, x, i)) \tag{23}$$

Where i represents the i-th input of the input variable x, y_i and y_{i-1} represent the current state and the previous state of the output sequence. t_k and s_l act as transition and state feature functions, respectively, both depending on the current position and usually taking values of 0 or 1. λ_k and μ_l are the corresponding weights for the feature functions.

By adding constraints to the labeled sequence, the CRF layer ensures that the final output sequence is valid. These constraints can be automatically learned by the CRF layer during the training process from the training dataset. The constraints may include: the first character of a sentence should start with "B-" or "O"; entity annotations should start with "B-", "O, I-label" is invalid; if a labeling sequence is "B-label1, O-label2, O-label3, ...", all labels in this sequence should belong to the same named entity. The constrained results obtained through constraint operations are then used to calculate the loss for the current sentence, evaluating the model's learning performance and influencing the backpropagation process.

3.6   Adversarial Training

There are two types of uncertainty in deep learning: aleatoric uncertainty and epistemic uncertainty. These uncertainties can cause fluctuations in model results and reduce the model's generalization ability. Aleatoric uncertainty typically stems from labeling errors in the dataset, and the more disordered the labeling noise, the greater the aleatoric uncertainty. Epistemic uncertainty, on the other hand, arises from observation errors caused by model parameter sensitivity, and a smaller training set exacerbates the impact of epistemic uncertainty on the model.

Adversarial training is a training method that enhances the model's generalization ability. It achieves this by adding certain perturbations to the original inputs to generate adversarial samples. These adversarial samples are then used to train the model, improving its ability to recognize adversarial examples and enhancing the model's generalization. Therefore, this paper attempts to introduce adversarial training mechanisms to mitigate the impact of these two uncertainties on the model and further improve the model's effectiveness.

This paper refers to the Fast Gradient Method (FGM) adversarial training mechanism [25]. For each embedded representation x of an input text sequence, a perturbation r_{adv} applied to the model is shown in Equations(24)and(25).

$$r_{adv} = \epsilon \cdot \frac{g}{\parallel g \parallel_2} \tag{24}$$

$$g = \nabla_x L(\theta, x, y) \tag{25}$$

Where x is the input sequence, y is the corresponding label for x, θ is the set of model parameters to be learned, and L is the loss function. Therefore, Equation(25) represents the gradient of the loss function with respect to the input sequence x. By calculating the gradient in this way, the embedded vector input is further moved in the direction where the loss gradient ascends the fastest, thus forming an adversarial attack. In Equation(24), the obtained gradient is normalized using the L2 norm. By applying such perturbations to the embedded representation, the model needs to search for more robust parameters during the training process to counter adversarial attacks and achieve better results.

The act of applying perturbations to the embedded representation of each input text sequence can, to some extent, simulate the labeling errors introduced by the annotation dataset itself. In the training process, finding more robust parameters can alleviate the impact of aleatoric uncertainty on the model to some degree. The influence of adversarial samples on the model also allows the model to tolerate changes brought about by parameter fluctuations, thereby mitigating the impact of epistemic uncertainty on the model.

4 Experiment Results and Analysis

4.1 Dataset

The training and validation datasets for our study were sourced from the annotated and unlabeled datasets provided by the CCKS 2017–2020 competition and the CHIP 2020 competition on medical named entity recognition. This paper combined four years of clinical data extracted from electronic medical records in the CCKS 2017–2020 competitions as experimental dataset 1, with 3,786 annotated data samples, surpassing a total of 1.32 million words, with an average sequence length of approximately 350 words per sample. Additionally, this paper obtained the medical text annotation data provided by the CHIP 2020 Chinese medical text named entity recognition competition as experimental dataset 2, with 20,000 annotated training samples totaling 2.2 million words, including 47,194 sentences. The datasets encompassed nine major medical entities, including 504 common pediatric diseases, 7,085 body parts, 12,907 clinical manifestations, and 4,354 medical procedures. As datasets 1 and 2 were annotated by different teams from different competition projects, combining them

Table 1. Introduction to CCKS2017-2020 and CHIP2020 Medical Named Entity Recognition Task Annotation Dataset

Source	Text Sample	Label Samples	Average Length	Dataset Size
CCKS 2017–2018	男、39岁，承德市双滦区人。主因上腹部、腰部疼痛1天入院。	上腹部 16 18 身体部位 腰部 20 21 身体部位 疼痛 22 23 症状和体征	220	1198
CCKS 2019–2020	{"originalText": "，患者于2012年12月无明显诱因出现右上腹隐痛，不剧可忍，伴粘液便，大便干结。","entities": [{"end_pos": 22, "label_type": "解剖部位","start_pos": 19}]}		410	2586
CHIP 2020	病程早期可闻胸膜摩擦音在全部呼吸期间均可听到。|||5 10 sym|||11 21 sym|||		54	20000

led to inconsistent rules resulting in ambiguous entity boundaries and inconsistent entity recognition and classification. Therefore, our experiment divided the datasets from the two competitions and used them separately as model inputs for training.

Table 1 presents the basic information of the datasets from the CCKS 2017–2018, CCKS 2019–2020, and CHIP 2020 competitions, including the sample text annotation format, average sentence length, and dataset sizes. The sample text for CCKS 2019–2020 had longer length, and the text was truncated for its content.

This experiment trained the models on the training set and evaluated the training effect using the validation set.

4.2   Evaluation Metrics

This paper adopted the same evaluation metrics as previous relevant literature, which used the harmonic mean of Precision and Recall (F1-score) to evaluate the model's performance. It describes the robustness of the model and is calculated as shown in Equations(26)–(28).

$$Precision = \frac{TP}{TP + FP} \tag{26}$$

$$Recall = \frac{TP}{TP + FN} \tag{27}$$

$$F1 = \frac{2 \times Precision \times Recall}{Precision + Recall} \tag{28}$$

where TP (True Positive) represents the number of samples that are actually positive and predicted to be also positive; FP (False Positive) represents the number of samples that are actually negative but predicted to be positive; FN (False Negative) represents the number of samples that are actually positive but predicted to be negative.

From the equations, it is easy to see that when either Precision or Recall has a small value, regardless of the other value being high, the F1-score will be small. F1-score only approaches 1 when both *Precision* and *Recall* tend towards 1. The F1-score calculation is independent of the number of positive and negative samples in the dataset. Therefore, in imbalanced prediction of samples with highly different proportions in different categories, the F1-score is more persuasive than accuracy.

4.3   Comparative Experiments

Baseline Model. To validate the effectiveness of the proposed model, this paper compared it with several models listed below. The description of each model is as follows:

1. BERT-BiLSTM-CRF: Baseline model that uses the basic Chinese pre-trained parameters of BERT to embed word vectors, BiLSTM for encoding, and CRF for labeling.
2. RoBERTa-BiLSTM-CRF: This model uses the improved Chinese pre-trained model RoBERTa's pre-trained parameters to embed word vectors, BiLSTM for encoding, and CRF for labeling.

Experimental Results and Analysis. This paper conducted experiments on the CCKS dataset using the proposed model, reproduced two commonly used model architectures in the named entity recognition field, and tested them on the dataset. Table 2 lists the test results of the models. From the table, it can be observed that the proposed model achieves better precision, recall, and F1-score than the existing models that use pre-trained models, demonstrating the advantages of the improved BERT model, hybrid encoding layer, and adversarial training mechanism.

Table 2. Test results of different models on the dataset

Model	Evaluation Metrics(%)			
	accuracy	precision	recall	F1-score
BERT-BiLSTM-CRF	98.00	89.84	**92.15**	90.97
RoBERTa-BiLSTM-CRF	98.00	91.13	92.64	91.86
Our model	98.00	**91.95**	92.86	**92.40**

4.4   Ablation Experiments

To demonstrate that the incorporation of the adversarial training mechanism, improved BERT model, and hybrid encoding layer indeed improves the model's performance, this paper conducted ablation experiments to validate the effectiveness of the proposed model structure, as shown in Table 3. (To simplify the expression, the first letters of the embedding layer, encoding layer, and normalization layer modules are concatenated as model abbreviations. 'Bert' will be simplified as 'B', 'RoBERTa' will be simplified as 'RB', 'BiLSTM' will also be simplified as 'B', 'R-Trans' will be simplified as 'R', 'CRF' will be simplified as 'C' and 'Adversarial training' will be simplified as 'A', thus, 'RoBERTa-BiLSTM-CRF with Adversarial learning' will be simplified as 'ARBBC'.)

Table 3. Results of ablation experiment with multiple models

Model	Evaluation Metrics(%)			
	accuracy	precision	recall	F1-score
BB&RC	98.00	90.82	91.23	91.01
RBB&RC	98.00	91.04	92.01	91.52
ABB&RC	98.00	91.33	92.00	91.65
ARBBC	98.00	91.66	**93.08**	92.36
ARBRC	98.00	91.34	92.74	92.03
Our method(ARBB&RC)	98.00	**91.95**	92.86	**92.40**

During the experiments, it was found that incorporating adversarial training and improved pre-training models can improve the accuracy of annotation. However, in the comparison of the ablation experiments APBBC and APBB&RC, it was found that the hybrid encoding layer does not have a significant impact on the overall F1-score of the experiments. Furthermore, the experiment analyzed and compared the F1-score for different entity classes between the RoBERTa-BiLSTM-CRF with Adversarial learning model and the proposed model, as shown in Table 4, with the data displayed in the format of (BiLSTM encodinghybrid encoding layer).

Table 4. Results of ablation experiment with multiple models

Entity Category	Evaluation Metrics(%)			Number of samples
	precision	recall	F1-score	
Treat	**83.24/88.24**	**83.24/86.54**	**83.24/87.38**	364
Operation	**91.43/91.88**	**91.43/92.14**	**91.43/92.01**	700
Imaging examination	**91.55/92.99**	**95.65/95.52**	**93.55/94.24**	736
Medicine	**95.71/95.60**	**95.78/96.52**	**95.75/96.06**	1352
Symptom and Sign	**96.31/97.31**	**96.72/96.77**	**96.51/97.04**	2319
Parts of Body	88.86/89.18	90.12/89.57	89.49/89.37	3230
Disease and Diagnosis	88.02/88.00	90.34/89.76	89.17/88.87	3407
Laboratory Test	95.21/95.18	96.05/95.76	95.63/95.46	3770
Anatomical Site	90.92/90.83	92.86/92.27	91.88/91.54	6676

The experimental results indicate that the hybrid encoding layer can effectively improve the recognition performance of entity classes with fewer samples, with a maximum improvement of 4.14%.

The results of the comprehensive comparison and ablation experiments indicate that the proposed APBB&RC model has better identification performance compared to the commonly used named entity recognition models, BBC and RBC. This validates the effectiveness of the APBB&RC model in incorporating adversarial training, improved BERT model, and hybrid encoding layers to obtain better model parameters, enhance word embeddings for expressing Chinese semantic information, and capture local and long-distance dependency information for generating more accurate text annotations.

5   Conclusion

This paper proposes a Chinese medical named entity recognition model based on RoBERTa (A Robustly Optimized BERT Pre-training Approach), adversarial training, and hybrid encoding layers. This model enhances the perception of semantic information and the generalization ability of the model. In this paper, we conduct research on three real medical case datasets to evaluate the proposed model. Experimental results demonstrate that the proposed model achieves significant performance improvement compared to the baseline models on real datasets. Furthermore, this paper analyze the advantages of this improvement over the baseline models through experiments.

References

1. Grishman, R., Sundheim, B.: Message understanding conference-6: a brief history. In: COLING, vol. 1 (1996)
2. Wu, Y.C., Fan, T.K., Lee, Y.S., Yen,S.J.: Extracting named entities using support vector machines. In: International Workshop on Knowledge Discovery in Life Science Literature, pp. 91–103 (2006)
3. Ju, Z., Wang, J., Zhu, F.: Named entity recognition from biomedical text using SVM. In: International Conference on Bioinformatics and Biomedical Engineering, pp. 1–4 (2011)
4. Zhou, G.D., Su, J.: Namedd entity recognition using an HMM-based chunk tagger. In: Meeting on Association for Computational Linguistics, pp. 473–480 (2002)
5. Mccallum, A., Li, W.: Early results for named entity recognition with conditional random fields, feature induction and web-enhanced lexicons. In: Conference on Natural Language Learning at HLT-AACL, pp. 188–191 (2003)
6. Mccallum, A., Freitag, D., Pereira, F.: Maximum entropy Markov models for information extraction and segmentation. In: Proceedings of the 17th International Conference on Machine Learning, pp. 591–598 (2000)
7. Feng, Y., Yingying, C., Gengui, Z., et al.: Intelligent recognition of named entities in electronic medical records. Chin. J. Biomed. Eng. **30**(2), 256–262 (2011)
8. Jianbo, L., et al.: A comprehensive study of named entity recognition in Chinese clinical text. J. Am. Med. Inform. Assoc. JAMIA **21**(5), 808–814 (2014)
9. Chen, H., Lin, Z., Ding, G., et al.: GRN: gated relation network to enhance convolutional neural network for named entity recognition. In: Proceedings of the 33rd AAAl Conference on Artificial Intelligence (2019)
10. Gehring, J., Auli, M., Grangier, D., et al.: Convolutional sequence to sequence learning. In: Proceedings of the 34th International Conference on Machine Learning-Volume 70-JMLR-org-s.libyc.nudt.edu.cn:443, pp. 1243–1252 (2017)

11. Mikolov, T., Karafiát, M., Burget, L., et al.: Recurrent neural network based language model. In: Interspeech, Conference of the International Speech Communication Association, Makuhari, Chiba, Japan, September. DBLP (2015). https://doi.org/10.1109/EIDWT.2013.25

12. Hochreiter, S., Schmidhuber, J.: Long short-term memory. Neural Comput. **9**(8), 1735–1780 (1997). https://doi.org/10.1162/neco.1997.9.8.1735

13. Wang, Q., Xia, Y., Zhou, Y., et al.: Incorporating dictionaries into deep neural networks for the Chinese clinical named entity recognition. J. Biomed. Inform. **92**, 103–133 (2019)

14. Xu, K., Zhou, Z., Hao, T., et al.: A bidirectional LSTM and conditional random fields approach to medical named entity recognition. In: Proceedings of International Conference on Advanced Intelligent Systems and Informatics, Cairo, Egypt, October 24–26, pp. 355–365 (2017)

15. Tang, B., Wang, X., Yan, J.: Entity recognition in Chinese clinical text using attention-based CNN-LSTM-CRF. BMC Med. Inform. Decis. Mak. **19**(S3), 74–82 (2019)

16. Devlin, J., Chang, M.W., Lee, K., et al.: Bert: pre-training of deep bidirectional transformers for language understanding. In: Proceedings of the 2019 Conference of the North American Chapter of the Association for Computational Linguistics: Human Language Technologies, Volume 1(Long and Short Papers), pp. 4171–4186 (2019)

17. Mikolov, T., Chen, K., Corrado, G., et al.: Efficient estimation of word representation in vector space. Comp. Sci. **2013**, 2–5 (2013)

18. Pennington, J., Socher, R., Manning, C.: Glove: global vectors for word representation. In: Conference on Empirical Methods in Natural Language Processing, pp. 1532–1543 (2014)

19. Dai, A.M, Le, Q.V.: Semi-Supervised Sequence Learning. MIT Press (2015)

20. Ramachandran, P., Liu, P.J., Le, Q.V.: Unsupervised Pretraining for Sequence to Sequence Learning (2017)

21. Peters, M., Ammar, W., Bhagavatula, C., et al.: Semi-supervised sequence tagging with bidirectional language models. **2017**, 2–9 (2017)

22. Radford, A., Narasimhan, K.: Improving Language Understanding by Generative Pre-Training (2018)

23. Liu, Y., Ott, M., Goyal, N., et al.: RoBERTa: a robustly optimized BERT pre-training approach. arXiv preprint arXiv:1907.11692 (2019)

24. Wang, Z., Ma, Y., Liu, Z., et al.: R-Transformer: Recurrent Neural Network Enhanced Transformer (2019). https://doi.org/10.48550/arXiv.1907.05572

25. Miyato, T., Dai, A.M., Goodfellow, I.: Adversarial training methods for semi-supervised text classification. In: International Conference on Learning Representations (2017)

A Function Fitting System Based on Genetic Algorithm

Qiuhong Sun[1(✉)], Jiaqi Wang[1,3], and Xiaokang Zhou[2,4]

[1] School of Information Science and Engineering, Hebei University of Science and Technology, Shijiazhuang 050018, China
sunqiuhong@hebust.edu.cn
[2] The Faculty of Data Science, Shiga University, Hikone 522-8522, Japan
zhou@biwako.shiga-u.ac.jp
[3] Department of Computer Science, North China Electric Power University, Baoding 071003, China
[4] The RIKEN Center for Advanced Intelligence Project, RIKEN, Tokyo 103-0027, Japan

Abstract. With the development of science and technology, function fitting has penetrated into various fields of scientific research, scientific and technological innovation. For the function of fitting analysis of a given function image, there is no public software on the market that uses the idea of genetic algorithm to solve the problem of function fitting. In order to make up for the insufficiency of the existing software and seize the opportunity of function fitting in various industries, a function fitting system based on genetic algorithm was proposed. It is a low-threshold software with a wide range of applications. When designing the function of fitting and analyzing a given function image, the genetic algorithm is used. With the application of ray detection in Unity, the closest expression of function fitting is obtained. At the same time, the line graph of fitness reduction during genetic algorithm iteration is given.

Keywords: Function Fitting · Genetic Algorithm · Natural selection

1 Introduction

1.1 Project Background

With the development of science and technology, function fitting has been deeply involved in various fields of scientific research and scientific innovation and manufacturing. There are many things cannot be separated from the support of function fitting, such as the evaluation and analysis of mechanical manufacturing, the surface recognition of sample material [1], the scientific research of crustal movement [2], the data analysis in the field of engineering economy [3] and so on.

This paper is funded by China Scholarship Council and the Provincial Master Ideological and Political Teaching Team Construction of Hebei Provincial Department of Education.

The research of function fitting theory is not just imaginary, but has historical origin. In some literatures, function fitting is divided into Ordered point cloud function fitting and Disordered point cloud function fitting [4]. It can be seen that artificial neural network is an important development tendency of function fitting in recent years. However, outstanding studies on function fitting using genetic algorithms have not been publicly described.

1.2 Market and Competition

The large global educated population provides a market basis for function fitting. The function fitting system based on genetic algorithm is easy to operate and suitable for people of any age. However, through demand analysis, it can be seen that the function fitting system is mainly used for students to complete their academic tasks, scientific researchers to engage in scientific research creation, and mathematical fans to carry out theoretical learning. Above all, the function fitting system will have a large application market [5].

Function fitting is an important branch of mathematical modeling, which is becoming more and more important in modern education [6]. In 2003, the Ministry of Education of China issued the Standard of Mathematics Curriculum for Ordinary High School (Experiment), which put forward the mathematical modeling as a component of the mathematics curriculum. In the Curriculum Standard (2017) issued by the Ministry of Education of China, mathematical modeling is a component of the six core qualities [7]. In addition, in recent years, there have been many scientific innovation competitions with mathematical modeling as the theme, which have attracted wide attention and participation of the society.

For the function of fitting analysis of a given function image, there is seldom public software on the market that uses the idea of genetic algorithm to solve the function fitting problem.

1.3 Project Content

Based on the related theory of genetic algorithm, the function of fitting analysis to the given function image is realized. It is easy to describe the function image according to the function expression, but it is opposite to find the function expression according to the function image. Grotesque objects are easy to be seen in our life, but not every contour of objects can be described by function expression. But genetic algorithm provides the possibility to fit almost everything in the world. It is a task of the function fitting system to analyze the given function image by using genetic algorithm.

2 Feasibility Study and Demand Analysis

2.1 Feasibility Study

It is highly feasible to apply the theory of genetic algorithm to the problem of function fitting. Genetic algorithm originated from Darwin's Natural selection theory; the main idea is survival of the fittest. It is a good theory to calculate the best solution of NP-hard problems. The fitting analysis of a given function image is not only a NP difficult problem, but also a problem of seeking optimal solutions. So genetic algorithm will be very suitable for fitting analysis of a given function image. In addition, experimental results show that genetic algorithm has advantages in global search and intrinsic heuristic random search, and can significantly improve the learning effect of adaptive fuzzy nervous system [8]. It can be concluded that genetic algorithm can be well adapted to the study of function fitting.

It is highly feasible to use Unity engine for genetic algorithm training. Previously, some creators have successfully designed software such as packing system [9], obstacle avoidance marching system and so on.

2.2 Requirement Analysis

In the function of fitting and analyzing the given function image, the user can obtain the approximate function expression according to the geometry in the scene, and present the effect of iterative evolution of genetic algorithm in the form of line graph.

3 Basic Theory

When developing the function fitting system, the Unity2018 development engine and Visual Studio2019 integrated development environment are used. The scene of fitting analysis for a given function image is set as the switchable two-dimensional vision.

In the genetic and evolutionary process of organisms, the existing structure is maintained through the idea of survival of the fittest, and iterative evolution is carried out through the theory of genetic variation [10]. On the basis of the existing initial population, individual evaluation is carried out to obtain the fitness of each individual. After selection, recombination and mutation of the population, the next generation is obtained. Each iteration of the population has a greater possibility to retain the gene combination that is more adaptable to the environment.

Based on the related theories of heredity and evolution in biology, genetic algorithm abstracts NP difficult problems in programming into a process of natural selection through the process of encoding and decoding [11]. The overall process of genetic algorithm used to solve practical problems is shown in Fig. 1.

It is a little hasty to apply genetic algorithm to solve function fitting problem directly. Compared with the function fitting problem, the traveling salesman problem is simple, clear and intuitive. In order to be more familiar with the practical application of genetic algorithm, the relevant knowledge of using genetic algorithm was first learned to try to solve the traveling salesman problem, and also the genetic algorithm to solve the obstacle avoidance problem of unmanned vehicles was roughly understood. The traveling salesman problem refers to the construction of roads between stations to ensure that all stations are connected while ensuring the shortest total distance. When the number of sites is small, it can be listed one by one. But when the number of sites is large, it is difficult to do exhaustive work. Genetic algorithm becomes a good solution. When genetic information is encoded, each station is coded as a natural number, so the travel route is regarded as an individual gene. When decoding genetic information, individual

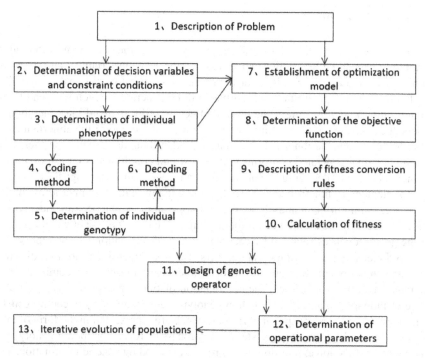

Fig. 1. The whole process of using genetic algorithm to solve practical problems

genes are converted into travel routes, and the entire journey is calculated as individual fitness. In the process of using genetic algorithm to solve the traveling salesman problem, the random contest method is used to design the selection operator, especially considering the individual fitness into the local optimal solution. In addition, the effect of the crossover operator is not to exchange alleles, but to pay attention to the position switch of the same genes while exchanging alleles, which solves the result of repeated gene recombination at the site.

In the basic steps of genetic algorithm, the process of optimization problem description, determination of decision variables, determination of constraints, and description of individual phenotype determines the way of encoding and decoding. After repeated theoretical research and programming practice, it is found that it is highly feasible to encode and decode genetic information through four operations of elementary function and mutual mapping of natural number array in this function fitting system. According to the previous study on the genetic algorithm to solve the traveling salesman problem, the difference between the global optimal solution and the local optimal solution is taken into consideration in the design of the selection operator. The optimal individual is given priority in the design of the crossover operator. The meaningless phenotype is especially noted in the design of the mutation operator.

4   Detailed Design

In terms of determining decision variables and constraints, the constraint of individual phenotype is a meaningful function expression, and the decision variable is an expression combination, including the combination of operands and operators. When the constraints of individual phenotype are not met, outliers will be generated, which will lead to the function fitting may not achieve the desired effect. In order to avoid the generation of outliers, the equivalent replacement of operators was carried out for the abnormal cases of 0 as the divisor and the abnormal cases that crossed the boundary, so that the phenotype of each individual generated conforms to the constraint conditions as far as possible.

The encoding and decoding of genetic information are realized through the four operations of elementary function and the mutual mapping of natural number array. In this software, 22 common elementary functions and 4 operators are mapped into independent natural numbers by the method of symbol coding in the theory of genetic algorithm, so that the gene sequences retained by each generation of population can be represented by a two-dimensional array of natural numbers. The first dimension represents different individuals in the population, and the second dimension represents the combination of individual function expressions. The combination of function expressions is the phenotype of individuals, and corresponding genotypes are obtained by encoding genetic information. Genotypes are specific symbols, which are represented in the form of integer values in the function fitting system. The process of reverse mapping genotype to combination of function expression is the process of decoding genetic information. The coding information after applying the genetic algorithm theory to the function fitting system is shown in Table 1.

Table 1. Genetic information coding table.

expression	identifier	value	expression	identifier	value
1	numberSelection	0	Math.Pow(2, x)	numberSelection	13
3	numberSelection	1	Math.Pow(3, x)	numberSelection	14
5	numberSelection	2	Math.Pow(0.5, x)	numberSelection	15
7	numberSelection	3	Math.Sin(x)	numberSelection	16
Math.PI	numberSelection	4	Math.Cos(x)	numberSelection	17
Math.Log(x, E)	numberSelection	5	Math.Tan(x)	numberSelection	18
Math.Log(x, 2)	numberSelection	6	Math.Asin(x)	numberSelection	19
Math.Log(x, 10)	numberSelection	7	Math.Acos(x)	numberSelection	20
x	numberSelection	8	Math.Atan(x)	numberSelection	21
Math.Sqrt(x)	numberSelection	9	+	operationSelection	0
x* x	numberSelection	10	−	operationSelection	1
x* x* x	numberSelection	11	*	operationSelection	2
Math.Exp(x)	numberSelection	12	/	operationSelection	3

Next, build the optimization model. Genetic algorithm involves the theory of biological evolution, which itself has great uncertainty. Moreover, function fitting is a huge problem, which further enlarges the uncertainty of genetic algorithm. Through programming practice, the corresponding approximate expression can be obtained by using traditional genetic algorithm to solve the function fitting problem, but the fitness attenuation curve in the iterative process cannot reach the expected effect, because the effect of the selection operator, crossover operator and mutation operator of traditional genetic algorithm has great potential uncertainty in the function fitting system. In this function fitting system, the selection operator is reformed, from the simple random contest method for natural selection, to use the random contest method for natural selection while preserving the optimal individual. Thus, the basic model of genetic algorithm is optimized in the function fitting system, and the expected iterative fitness attenuation curve is obtained.

Fitness evaluation is an important step for genetic algorithm to solve practical problems. In the function fitting system, the purpose of function fitting analysis is to find the function curve closest to the image. Therefore, in two-dimensional image, the area between the image and the curve is taken as the parameter of fitness evaluation. In 3D image, the volume between image and hook face is used as the parameter of fitness evaluation. Programming can only solve the limited problem, so through the solution idea of definite integral, the calculation of the value of area or volume is transformed into the accumulative calculation of the area of a finite curved trapezoid or the volume of a finite curved top cylinder. Among them, the smaller the cumulative value is, the closer the fitting curve and function image are [12].

The primary population is generated by random numbers. The randomly generated expression is not necessarily an expression with operational significance. For the problem that 0 is used as a divisor, we replace the operator of the same level by detecting, that is, we change the operand 0 into 1 when we change the division sign of the operator into the multiplication sign, so as to ensure that the randomly generated expression has operational significance. In this software, the first half of the individual sequence of the population stores operands and the second half stores operators. Each randomly generated individual has a gene sequence proportional to the complexity of the function expression. The cumulative fitness of each individual is realized by asynchronous function, so the cumulative fitness evaluation is delayed when this software is designed.

Genetic operators include three types: selection operators, crossover operators and mutation operators. Designing genetic operators is the core content of genetic algorithms. Selecting appropriate genetic operator design methods will help improve the accuracy and efficiency of function fitting system.

The design of selection operator adopts the random contest method. All the individuals were divided into 8 groups in the form of 10 individuals in each group. According to fitness evaluation, half of the individuals in each group that were most adapted to the environment were selected as the winners. The winners are selected as the result of the genetic algorithm selection operator. The individuals selected by the random contest method in the function fitting system are representative. The method fully considers the correspondence between the local optimal solution and the global optimal solution.

The 40 individuals selected by the random contest method will make a copy of themselves. The original individuals of the 40 winning individuals will undergo subsequent gene recombination and gene mutation operations, and the backup 40 individuals will be directly retained to the next generation as the winning individuals, ensuring that the overall trend of the evolution of the group will develop in a good direction. Among them, the process of selecting winners based on fitness evaluation uses the idea of bubble sorting in data structure.

The crossover operator is designed with the single point crossover method. According to the result of natural selection by random contest method, the original individuals of 40 winning individuals were randomly mated. In the process of population iteration, gene recombination will occur with a specified probability. In this software, the single point crossover method was adopted to simulate gene recombination. After the individuals of the population were randomly shuffled, the random mating process was simulated by traversing the whole population with the rule of step size of 2. In the process of random mating, the starting position of gene recombination is randomly generated, and the gene in this position and the genes after it are genetically recombined with the corresponding individuals. After traversing all individuals, the result of the crossover operator of genetic algorithm is generated.

The mutation operator is designed with the basic bit mutation method. After natural selection by random contest method, the winners will have gene mutation at a specified probability during random mating. In this software, the basic bit mutation method is adopted to simulate gene mutation: the mutation position of gene is randomly generated, and the gene corresponding to two phenotypes of operand and operator fitted by function is mutated. After traversing all individuals, the result of mutation operator of genetic algorithm is generated. The variation process of natural selection itself has great uncertainty, so the basic bit variation method is convenient for the design of the function fitting system, and can make a full attempt on biological evolution.

Determine the running parameters. The running parameters to be selected in genetic algorithm mainly include the length of individual coding string, the size of population, the probability of recombination, the probability of mutation, the number of iterations and so on. The length of individual coded string is represented by the complexity of function expression in the function fitting system, and the length is set as 10. In this function fitting system, the population size is represented by the number of randomly generated function expressions in each generation. In order to improve the efficiency of the program as much as possible on the premise of reflecting the diversity of the group, the group size is set at 80. The meanings of the probability of recombination, the probability of mutation and the number of iterations is consistent with those described in the classical genetic algorithm. In order to carry out iterative evolution on the premise of preserving excellent individuals as much as possible, the recombination probability and mutation probability of the function fitting system were set as 0.5 and 0.1 respectively. In order to shorten the program running time as much as possible on the premise of obtaining a good fitness attenuation curve, the number of iterations was set at 200.

So far, according to the basic theory of genetic algorithm, we have completed the relevant programming design. Then iteration is carried out by recursion method, and the best fitness of each generation of individuals is traced, forming the fitness attenuation curve of the iteration of genetic algorithm, so as to clearly show the effect of 200 generations of evolution of 80 individuals within the population.

Because the genetic algorithm has a long calculation process and the hardware configuration of each user's machine is different, there is no guarantee that every machine can get the result instantaneously. Therefore, when using the function of fitting analysis on the image of a given function, it is necessary to wait patiently for the results of the genetic algorithm.

Because of the great uncertainty of genetic algorithm, it cannot guarantee that it can get a good fitting function. The function fitting system sets the number of population individuals as 80 and the number of iterations as 200, which ensures that the time of genetic algorithm calculation is as short as possible while the function fitting is as effective as possible.

Before function fitting by genetic algorithm, the initial interface contains only the original function image, ray detection track and necessary operation prompts. After all calculations of the genetic algorithm are completed, the best fitting curve will be generated, and the best fitting expression will be modified accordingly, as shown in Fig. 2.

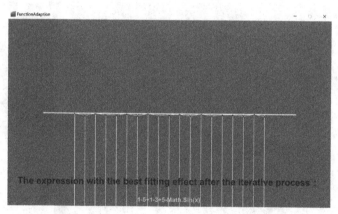

Fig. 2. Perform the best expression test for horizontal line fitting

Click the keyboard Q to rotate the field of view to the rear, so that the fitness attenuation curve in the genetic algorithm can be acquire, as shown in Fig. 3.

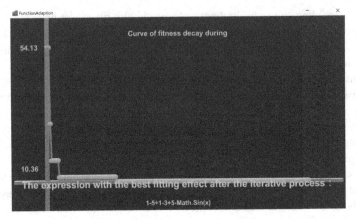

Fig. 3. Fitness attenuation curve of iterative process when horizontal fitting is performed

In the scene where fitting cases are selected, click "diagonal fitting" to carry out the process similar to the above "horizontal fitting", and the corresponding function image fitting effect is shown in Fig. 4.

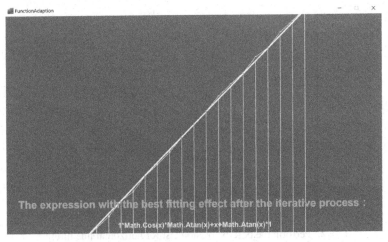

Fig. 4. Perform the best expression test for diagonal fitting

The fitness attenuation curve of "diagonal fitting" obtained by iteration of genetic algorithm is shown in Fig. 5.

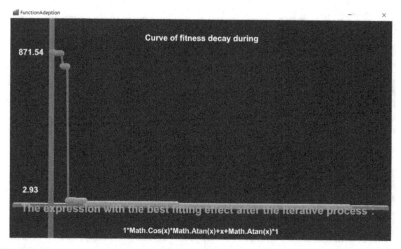

Fig. 5. Fitness attenuation curve of iterative process when diagonal fitting is performed

In the scene of selecting fitting cases, click "parabolic fitting" to carry out the process similar to the above two functions. The image fitting effect of corresponding functions is shown in Fig. 6.

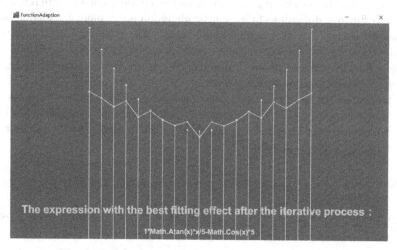

Fig. 6. Perform the best expression test for parabolic fitting

The fitness attenuation curve of "parabolic fitting" obtained by iteration of genetic algorithm is shown in Fig. 7.

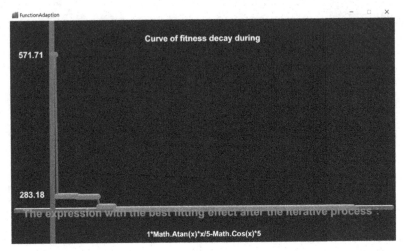

Fig. 7. Fitness attenuation curve of iterative process when parabolic fitting is performed

5   System Test

5.1   The Result of Tests

The iterative calculation time of genetic algorithm will be different with different graphs. The graph which is difficult to fit by genetic algorithm needs more calculation, so the calculation time will be longer. With the hardware support of CPU (i7-9750H) and GPU (GTX1660Ti), the average running time of the function fitting task of 80 individual iterations for 200 times is about 2 min and 20 s when the function fitting analysis function of a given function image is realized.

According to the theory of equivalence partitioning and boundary value analysis of software engineering [13], multiple groups of test cases are used to test the fitting analysis function of the image of a given function. The running process is smooth, and the test results are shown in Table 2.

Table 2. Black box test results for the function of fitting analysis for a given function image

id	test content	test cases	test times	success rate
01	Function fitting of geometric figures	horizontal fitting	20	65%
02	Function fitting of geometric figures	diagonal fitting	10	100%
03	Function fitting of geometric figures	parabolic fitting	10	100%
04	Function fitting of geometric figures	grotesque fitting	10	100%

Through the analysis of the test results, it can be seen that the fitting analysis function of the given function image of the system meets the design requirements, and also shows the uncertainty of the evolution of genetic algorithm.

5.2 Evaluation of Tests

At present, the function fitting system based on genetic algorithm is version 1.0, completing all the contents in the process of problem definition and achieving a friendly human-computer interaction effect. The next step will focus on improving the success rate of some test cases in the function of fitting analysis on the given function image and improving the corresponding algorithm. In order to improve the success rate of function fitting, the alternate scheme of Taylor expansion method will be used for the second function fitting. According to the theory of higher mathematics, we can carry out Taylor expansion of elementary function expression [14], that is, transform it into polynomial form. Therefore, in the stage of genetic information coding, the function fitting based on genetic algorithm is completed by setting the mutual mapping between the power function and the integer type. When the previously designed elementary function expression fails to fit, the alternative function fitting scheme based on Taylor expansion method is started to obtain better results.

After improving the success rate of some test cases in the fitting analysis function of a given function image, the system functions will be improved in three stages: initial stage, middle stage and later stage. In the initial stage (within 1 year), the function of training record storage is added to provide reference for the evolution of genetic algorithm. In the middle stage (1–3 years), better coding methods of genetic information are explored and more accurate function fitting is carried out. In the later stage (after 3 years), improve the network content and realize the sharing of system data.

6 Conclusion

As a piece of software, the function fitting system aims to solve the problem of function fitting with a new way of thinking. When designing the function of fitting analysis for a given function image, the relevant theory of genetic algorithm is used. The optimal expression can be generated after the iterative calculation of genetic algorithm for the selected function graph, and the fitness attenuation of the iterative process of genetic algorithm can be viewed after switching the vision.

References

1. Meng, P.: Application of Origin custom function fitting in the analysis of soil adsorption isothermal model. Exp. Technol. Manage. **34**(1), 7 (2017)
2. Liu, Q., Sun, G.: Application of multihedral function fitting in the study of horizontal crustal movement. Earthq. Res. Shanxi **1**(4), 6 (1996)
3. Zhang, X.: Application of quadratic function fitting in the field of engineering technology and economy. Fujian Arch. **1**(5), 3 (2009)
4. Chen, W.: Research on Curve Fitting Principle and Its Application. Changsha University of Science and Technology, Changsha (2018)
5. Tian, T.: Research on the Development of High School Students' Mathematical Modeling Literacy under the Background of "Internet Plus" Education. Ningxia Normal University, Guyuan (2020)

6. Lv, M., Ma, L.: The development trend, problems and countermeasures of the integration of digital economy and real economy in China during the 14th five-year plan period. Decis. Inf. **1**(2), 6 (2002)

7. Song, H.: Investigation and Research on the Status Quo of High School Students' Mathematical Modeling Literacy. Hebei Normal University, Shijiazhuang (2019)

8. Ding, H., Liu, S.: Application of ANFIS in function fitting based on GA. Microcomput. Inf. **25**(3), 3 (2009)

9. Zhao, Z., Chen, Q., Li, T., Zhang, J.: Design of unity3d packing system based on genetic algorithm. Comput. Modern. **1**(6), 5 (2014)

10. Zhou, M., Sun, S.: Principle and Application of Genetic Algorithm, pp. 4–6. National Defense Industry Press, Beijing (1999)

11. Karr, C.L., Weck, B., Massart, D.L., et al.: Least median squares curve fitting using a genetic algorithm. Eng. Appl. Artif. Intell. **8**(2), 177–189 (1995)

12. Department of Mathematics, Tongji University: Advanced Mathematics, 6th edn. Higher Education Press, Beijing, pp. 202–217 (2007). (in Chinese)

13. Zhang, H.: Introduction to Software Engineering, 4th edn., pp. 100–130. Tsinghua University Press, Beijing (2003)

14. Pu, J., Duan, X., Ye, Y., Chen, X.: Progressive construction and explicit calculation of multi-point Taylor expansion and its application. J. Hangzhou Dianzi Univ. Nat. Sci. Ed. **34**(5), 4 (2014)

A Rumor Detection Model Fused with User Feature Information

Wenqian Shang[1], Kang Song[1], Yong Zhang[2], Tong Yi[3(✉)], and Xuan Wang[1]

[1] State Key Laboratory of Media Convergence and Communication, Communication University of China, Beijing 100024, China
[2] SD Steel Rizhao Co., Ltd, Rizhao 276825, China
[3] School of Computer Science and Engineering, Guangxi Normal University, Guilin 541004, China
yitong@mailbox.gxnu.edu.cn

Abstract. With the rapid development of artificial intelligence technology, people's communication become more frequent, enjoying convenient at the same time, also aggravated the spread of rumors and spread. Therefore, rumor detection in social platforms has become an important direction of current scientific research. From the perspective of User characteristics, this paper uses deep learning methods to mine the change trend of user characteristics related to rumor events, and designs a rumor detection Model (User Feature Information Model, UFIM). Firstly, the feature enhancement function is used to recalculate the user feature vector to obtain a new feature vector representing the user's comprehensive feature under the current event. Then, the GRU model and the CNN model are used to learn the global and local changes of user features with the development of the event, and the user and time information are used to learn the hidden rumor features in the process of rumor spreading. The experimental results show that the UFIM model improved performance compared with the baseline model, rumors can effectively realize detection task.

Keywords: Rumor Detection · User Characteristics · CNN · GRU

1 Introduction

Rumor detection had received widespread attention in recent years and many related methods had been proposed. Most of the previous rumor detection methods regard rumor detection as a binary classification problem in supervised learning. That is, the constructed classifier can classify the targeted information into rumor or non-rumor labels. Existing rumor features are mainly selected from the text content, user attributes, and information dissemination. The previous rumor detection methods can be divided into two categories: rumor detection based on traditional machine learning and rumor detection based on deep learning.

We leverage deep learning models to study the change of user characteristics in the process of spreading trend. The proposed method can recapture user characteristics to adapt to the needs of the current topic in the specific scenario of rumor detection. The main contributions of this paper can be summarized as follows.

H. Jin et al. (Eds.): GPC 2023, LNCS 14503, pp. 169–176, 2024.
https://doi.org/10.1007/978-981-99-9893-7_13

(1) In order to solve the problem that user characteristics will shift with the development of the event in the process of event propagation, this paper designs a rumor detection model called UFIM. Firstly, the proposed model uses the feature enhancement function to recalculate the user feature vector, and obtains a new feature vector that can represent the user's comprehensive characteristics under the current event. Then the GRU model and CNN model are leveraged to learn the global and local changes of user characteristics with the development of events. In addition, the UFIM model can perform efficient and stable detection in the early stage of propagation.
(2) The experiments are conducted on the real-world dataset. The experimental results indicate that UFIM outperforms the baseline mode, and the practice of the proposed model is verified.

2 Related Work

There are three categories of features that can be used to distinguish between rumors and non-rumors in rumor detection: text features, user features and propagation features. Early research mostly extracted features from news text content to detect the authenticity of news articles. The text content includes the original post content and the user's response. For example, Castillo et al. 7 regarded the proportion of positive blog posts, the proportion of negative blog posts, and the proportion of blog posts including the first personal pronoun as text features, and adopted C4.5 classifiers to identify false news topics on Twitter. Yang et al. 8 considered link URLs, the number of positive and negative emotional words, while analyzing the text features. Then the SVM classifier is utilized to achieve rumor detection tasks. Ma et al. 5 leveraged Recurrent Neural Network to learn user response characteristics by useful query phrase of user response. The rumor detection methods in literature [4–6] focus on the language style, writing style and social emotion, respectively.

Based on the forementioned research work, this paper proposes a rumor detection model, called UFIM, which considers the user deep features and temporal signal changes during user information transmission. The proposed method can improve the performance in rumor detection.

3 UFIM Model

3.1 Definition of Problem

The rumor detection model proposed in this paper belongs to a two-category task, which is to judge whether the event is a rumor or a non-rumor based on the sequence of users participating in the event propagation (users are ranked according to the chronological order of users' comments). The objective function is as follows.

$$\hat{y} = f(X) \tag{1}$$

where $\hat{y} \in \{0, 1\}$, 0 and 1 denotes that the event is true and is a rumor, respectively.

3.2 Model Architecture

UFIM is mainly composed of three modules: data preprocessing module, feature extraction module, and feature fusion and classification module.

3.3 Data Preprocessing

The overall structure of the data preprocessing module mainly implements two tasks: user feature vector representation and user time series modeling. Both the two tasks would be elaborated in this subsection.

The Representation of User's Feature Vector

The main purpose of representation for user's feature vector is to represent the user's basic information in the form of vectors and obtain the user's initial features, this task is very helpful for the subsequent training and prediction of the model.

First of all, collect the user data required by the model, which includes the name of the account, personal information introduction, number of followers, number of fans, number of historical posts, user authentication information and whether the geographical location is enabled.

Then, different types of user information are preprocessed to obtain the corresponding vector representation of this type of information, and the processing method is shown in Table 1. Finally, the vector representation of all types of information is spliced to obtain the user feature vector representation.

Table 1. Table of preprocessing methods

No	Preprocessing mode	Type	User's information
1	String length representation method	Char	Personal Information
2	Numerical representation method	Int	Number of followers, number of fans, number of posts
3	Time representation method	Date	Account age
4	Boolean value representation method	Boolean	Whether to authenticate and enable geolocation

User Time Series Modeling

To learn the change of user characteristics with time in the process of event propagation, the construction of user time series is needed. To form a feature matrix, it arranges user feature vectors according to the time sequence of user comments.

3.4 Feature Extraction

The feature extraction module mainly learns the potential time difference features of users. This module is mainly composed of three parts: feature enhancement, local feature extraction and global feature extraction, which will be elaborated as follows.

Feature Enhancement

In order to represent the user characteristics under specific events more accurately and achieve the purpose of enhancing the original user characteristics, this paper designs a feature enhancement function. The calculation formula of the feature enhancement function is shown in Eq. (2).

$$f(t) = e^{-(\sigma(\sum_{k=1}^{n} w_{ik}x_{kj} + b_{ij}) + \beta)t} \tag{2}$$

where t is the time span that denotes the difference between the time of the user participation and event occurrence, i represents the user Posting order, x_{ij} represents the jth feature of the ith post user. $W_0 \epsilon R^{n \times n}$ and $B_0 \epsilon R^{n \times d}$ are initial weight matrices and bias matrices that are randomly set, where there are $w_{ik} \epsilon W_0$, $b_{ij} \epsilon B_0$. $\sigma(\bullet)$ represents the activation function, and the ReLU function is chosen as the activation function to avoid negative values. β is the fixed deviation value, giving a biased upper bound on the final weight.

The feature matrix X is processed by the feature enhancement function to obtain the attention weight matrix V. The aging feature matrix X' is obtained after matrix dot multiplication of the attention matrix V and the feature matrix X.

We use the balance coefficient to combine the two matrices of the original feature matrix and the aging feature matrix. Then the feature matrix X_{final} is obtained, the user characteristics can be represented by it. The formula for calculating X_{final} is shown in Eq. (3), where φ is the equilibrium factor.

$$X_{final} = X + \varphi \cdot X' \tag{3}$$

Local Feature Extraction

CNN is leveraged to learn the local change information of the user time series. The CNN model structure is shown in Fig. 1.

The user feature matrix X_{final} enhanced by feature enhancement function is fed to the convolutional neural network.

Global Feature Extraction

Recurrent Neural Network (RNN) is a kind of neural network model specially used to process sequence data. Therefore, the proposed method uses GRU model to learn the temporal relationship and hidden features of user time series, and realize the extraction of user global features in the process of event propagation. Figure 2 shows the structure of GRU model.

$x_t (t = 1, 2, \ldots, n)$ represents the input data of the GRU mode in Fig. 2, where there are two inputs of the GRU unit: the user aging feature x_t corresponding to the GRU unit and the output state h_{t-1} at the last time. The GRU model first takes the user time series

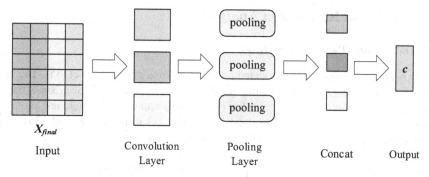

Fig. 1. The structure of CNN

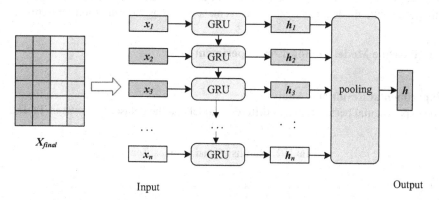

Fig. 2. Structure of GRU

in the matrix X_{final} as the input data of the model, and then uses the GRU unit to learn the input sequence data. Finally, it performs the mean pooling operation on these output states $h_t (t = 1, 2, \ldots, n)$ to obtain the user global time series feature representation h. The calculation method is shown in Eq. (4).

$$h = \frac{1}{n} \sum_{t=1}^{n} h_t \qquad (4)$$

Feature Fusion and Classification

In the rumor detection task, both local and global features are very important sources of information. Therefore, in order to better realize the task of rumor detection classification, local features and global features need to be fused and classified. A common approach is to concatenate the representation vectors of the two types of features together, and meanwhile further process and classify them by fully connected layers and activation functions. The advantages of the two features can be fully utilized to enhance the robustness and accuracy of the model with this method, hence the performance of the rumor detection would be improved.

4 Experiment Results and Analysis

4.1 Dataset and Parameter Settings

The experiment evaluation is conducted on the microblog public data set [35]. The data set includes all user-related information used in the model, such as the user's nickname, personal profile, number of fans, number of followers and so on 5. There are a total of 4664 events in the dataset, among which there are 2313 rumor events and 2351 non-rumor events. The total number of users is 2,746,818, and the total number of comments is 3,805,656.

In this paper, rumors are used as positive samples and the events in the dataset are divided according to the ratio of 7:2:1. Specifically, in this chapter, 70% of the events are used as the training set to train the model, 20% of the events are used as the test set 1 to evaluate the model, and 10% of the events are used as the validation set to tune.

4.2 Baseline Model Comparison Experiment

Experimental Result and Analysis

The experimental performance of different models on the dataset is shown in Table 2.

Table 2. Results of model comparison

Model	Category	Acc	P	R	F1
DTC[1]	T	0.831	0.815	0.847	0.830
	F		0.847	0.815	0.831
SVM-RBF[7]	T	0.818	0.815	0.824	0.819
	F		0.822	0.812	0.817
SVM-TS[3]	T	0.857	0.878	0.830	0.857
	F		0.839	0.885	0.861
DTR[8]	T	0.732	0.726	0.749	0.737
	F		0.738	0.715	0.726
GRU[9]	T	0.910	0.952	0.864	0.906
	F		0.876	0.956	0.914
RFC[10]	T	0.849	0.947	0.739	0.830
	F		0.786	0.959	0.864
PPC[11]	T	0.916	0.955	0.876	0.913
	F		0.884	0.957	0.919
BTUFC	T	**0.929**	**0.945**	**0.914**	**0.929**
	F		**0.932**	**0.924**	**0.928**

Observing the experimental results in Table 2, the experimental results of the proposed UFIM on microblog data are better than the baseline model, and the accuracy of the model reaches 92.9%, which verifies the effectiveness of the UFIM in the rumor detection task. It is noteworthy that both our model and PPC are grounded on CNN and GRU, and the proposed model outperforms PPC. The superiority of our model indicates that the user timeliness feature proposed in this paper can achieve the effect of enhancing user characteristics.

In summary, the experimental results indicate that UFIM has a good performance in both the accuracy and early rumor detection tasks.

5 Conclusion

The user characteristics used in the previous rumor detection models may be biased, because there are the differences among event topics. To address this matter, this paper designs a rumor detection model, called UFIM. The proposed model first uses the feature enhancement function to generate a user feature vector that can more accurately represent and highlight the association and difference between users, and then leverages GRU and CNN to extract the global and local changes in time of the feature, respectively. The experimental results indicate that the performance of UFIM is improved when compared with the baseline models. The proposed model also has its limitation. This paper mainly realizes the rumor classification task by learning the user characteristics under a single event. That is, the user set under an event is considered to exist in isolation, and the coupling and correlation between events and events are ignored. In future research, the co-occurrence relationship between users under rumor events and real events can be simulated to further improve the performance of the rumor detection model.

Acknowledgement. This work is partly supported by "the Fundamental Research Funds for the Central Universities CUC230A013".

References

1. Kwon, S., Cha, M., Jung, K.: Rumor detection over varying time windows. PloS one **12**(1), e0168344 (2017). https://doi.org/10.1371/journal.pone.0168344
2. Ma, J., Gao, W., Wei, Z., Lu, Y., Wong, K.F.: Detect rumors using time series of social context information on microblogging websites. In: Proceedings of the 24th ACM International on Conference on Information and Knowledge Management, pp. 1751–1754 (2015)
3. Yang, F., Liu, Y., Yu, X., Yang, M.: Automatic detection of rumor on Sina Weibo. In: Proceedings of the ACM SIGKDD Workshop on Mining Data Semantics, pp. 1–7 (2012)
4. Kwon, S., Cha, M., Jung, K., Chen, W., Wang, Y.: Prominent features of rumor propagation in online social media. In: 2013 IEEE 13th International Conference on Data Mining, pp. 1103–1108. IEEE (2013)
5. Ma, J., et al.: Detecting rumors from microblogs with recurrent neural networks (2016)
6. Ma, J., Gao, W., Wong, K.F.: Rumor detection on twitter with tree-structured recursive neural networks. Assoc. Comput. Linguist. (2018)
7. Castillo, C., Mendoza, M., Poblete, B.: Information credibility on twitter. In Proceedings of the 20th International Conference on World Wide Web, pp. 675–684 (2011)

8. Popat, K.: Assessing the credibility of claims on the web. In: Proceedings of the 26th International Conference on World Wide Web Companion, pp. 735–739 (2017)
9. Potthast, M., Kiesel, J., Reinartz, K., Bevendorff, J., Stein, B.: A stylometric inquiry into hyperpartisan and fake news. arXiv preprint arXiv:1702.05638 (2017)
10. Guo, C., Cao, J., Zhang, X., Shu, K., Yu, M.: Exploiting emotions for fake news detection on social media. arXiv preprint arXiv:1903.01728 (2019)
11. Liu, Y., Wu, Y.F.: Early detection of fake news on social media through propagation path classification with recurrent and convolutional networks. In: Proceedings of the AAAI Conference on Artificial Intelligence, vol. 32, issue 1 (2018)

Design and Implementation of a Green Credit Risk Control Model Based on SecureBoost and Improved-TCA Algorithm

Maoguang Wang[1,2], Jiaqi Yan[2]($\boxtimes$), and Yuxiao Chen[2]

[1] Engineering Research Center of State Financial Security, Ministry of Education, Beijing, China

[2] Central University of Finance and Economics, Beijing 102206, China
yjq_cufe@163.com

Abstract. Green credit plays a crucial role in promoting green transformation of enterprises and advancing social sustainable development. However, the current green credit rating disclosure system lacks data sharing between different institutions, leading to inconsistencies in evaluation results. To address this issue, this study proposes a green credit risk control model based on SecureBoost and an Improved-TCA algorithm. The proposed model combines vertical federated learning result with feature transfer to protect the privacy of participants in different datasets and analyzing the experimental results of vertical federated learning using SHAP values. We proposes improved TCA, which combines the BDA algorithm with the TCA algorithm, and improves the TCA algorithm by setting different weight ratios to comprehensively integrate the advantages of both algorithms to address the issue of significantly different sample distribution quantities in certain data set applications. We proved that the improved TCA algorithm combined with secureBoost has a better prediction result in the multi-classification credit evaluation scenario.

Keywords: Green Credit · SecureBoost · Improved TCA

1 Introduction and Motivation

Since the establishment of the world's first environmental protection bank "Ecology Bank" in 1974, green credit [1] has been developed for nearly half a century and has formed the basic principles of green credit business – "Equator Principles (EP)". Several scholars have empirically demonstrated the positive effects of green credit. For example, Eshet (2017) found that green credit services can enhance the competitiveness of enterprises through a thorough study of the impact of green credit services on enterprises. Zhao Qingxiang (2023) [2] empirically demonstrated that the implementation of green credit by banks can help reduce their risk exposure while improving the bank's operational performance. Scholars such as Shu Limin and Liao Jinghua (2023) have shown that green credit policies can promote corporate green innovation [3].

H. Jin et al. (Eds.): GPC 2023, LNCS 14503, pp. 177–191, 2024.
https://doi.org/10.1007/978-981-99-9893-7_14

However, current research in the field of green credit is mostly focused on empirical and policy analysis, and there has little research on risk control models in the field of green credit in recent years. Since the rating objects among various institutions are real listed companies, the convergence of sample ID among various institutions is higher. However, the feature dimensions and feature partition criteria among various data sets are different. Moreover, the rating basis of green credit rating agencies mostly originates from self-collected information of the institution and public disclosure of annual reports and social responsibility reports of listed companies, which have problems such as incomplete information collection and inadequate awareness of potential risks. Furthermore, the evaluation is non-official and has certain biases and hidden dangers. The distribution of sample numbers between actual enterprise green credit rating levels is uneven, showing a trend of normal distribution as a whole. Traditional transfer learning algorithms are difficult to present good predictions for all levels of future enterprise green credit classification based on historical data, and banks and consumers are also unable to predict enterprise green credit ratings using existing indicators.

To address the aforementioned issue, this study employs a feature-based transfer learning algorithm [4] and combines it with vertical federated learning [5] to enhance the TCA algorithm. Specifically, we integrate the classic feature-based transfer learning TCA [6] algorithm and BDA [7] algorithm to devise an improved TCA algorithm that weights the results of both algorithms based on the distribution of sample labels. Furthermore, we utilize the SecureBoost [8] algorithm and apply vertical federated learning to expand and select feature indicators of the experimental dataset. We also adopt SHAP values [9] to analyze experimental results and perform feature selection. The resulting improved TCA algorithm is used as the final transfer training model. The experimental results demonstrate that the proposed algorithm outperforms XGBoost [10], Random Forest, TCA, and BDA in the field of green credit risk control transfer training. Compared to TCA, Improve-TCA improves ACC by 10.73%, Precision by 20.67%, and Recall by 5.90%.

This paper will introduce the relevant work and research on feature transfer algorithms in the field of green credit risk control in Sect. 2, Sect. 3 will present the proposed feature transfer model based on the improved TCA algorithm and the related algorithms used in this paper. Section 4 will describe the experimental dataset and the experimental results adopted in this paper. Section 5 will present the corresponding conclusions obtained in this study. Lastly, Sect. 6 will discuss the future work plans for this study.

2 Related Work

With the continuous development of green credit in recent years, research on green credit has also been steadily increasing. However, currently, scholars mainly focus on the impact of green credit on the economy and corporate transformation, and little attention has been paid to the model construction in the field of green credit risk control. This paper will address this issue by combining the vertical federated learning SecureBoost algorithm with the feature transfer model based on the improved TCA algorithm.

The main idea of feature-based transfer learning [11] is to learn a pair of mapping functions $\{\varphi_s(\cdot), \varphi_t(\cdot)\}$, and map data from the source domain and target domain to the same feature space, reducing the differences between the domains. After training on the mapped data, a new target classifier is obtained. For unseen data in the target domain, it needs to be mapped to the new feature space and classified using the target classifier that has already been trained.

TCA and BDA are two classic algorithms in feature-based transfer learning. The main idea of TCA is to minimize the inter-domain kernel distance by imposing certain constraints. Based on TCA, BDA also considers the optimal distribution of classes between the source and target domains. By incorporating corresponding constraints, BDA optimizes both the inter-domain kernel distance and the distribution of classes across domains.

Vertical federated learning aligns overlapped data samples among different datasets and extends the corresponding features of the data samples to enhance training effectiveness. Boosting algorithm, as a typical algorithm of ensemble learning, has the core idea of combining the prediction results of multiple weak classifiers to construct a stronger classifier and achieve better results.

SecureBoost is an extended version of the Boosting algorithm, which extends the Boosting algorithm to federated learning tasks and provides some analysis and proof on privacy protection. The SecureBoost algorithm has been shown to achieve the same accuracy as the same algorithm without privacy protection in a non-federated setting.

SHAP (SHapley Additive exPlanations) is a method for explaining machine learning models. Its core idea is to calculate the marginal contribution of each feature to the model output and provide corresponding interpretation of the feature contributions to the model. The SHAP model obtains the SHAP value of each feature by calculating the output of each instance of the model and its influence on all features.

3 Green Credit Risk Control Model Based on SecureBoost and Improved TCA

The proposed model in this paper consists of two main parts. (see Fig. 1).The first part is the vertical federated learning component, which uses SecureBoost algorithm to perform vertical federated learning on different datasets from the same year, and selects the corresponding features to form a new dataset. The second part is the feature-based transfer learning component, where we propose a feature transfer model based on the improved TCA algorithm. In this component, we perform transfer predictions on datasets from different years and obtain corresponding results.

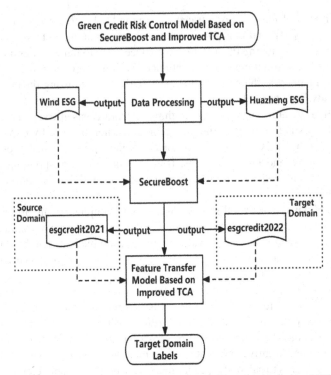

Fig. 1. Green Credit Risk Control Model based on SecureBoost and Improved TCA Algorithm.

3.1 Subsection SecureBoost Algorithm

SecureBoost (Secure federated tree-boosting, SecureBoost) algorithm, as a classic algorithm in vertical federated learning, aims to align samples with the same sample ID among different participants under privacy protection, and enables all participants to learn through the SecureBoost algorithm via privacy-preserving protocol [12].

The SecureBoost algorithm divides the participants into active and passive parties, with the active party providing sample labels and features and the passive party only providing sample features. Homomorphic encryption and decryption techniques are used to transmit corresponding data and parameters between active and passive parties to ensure data confidentiality. The active party leads the overall algorithm training, while the passive party provides the corresponding features and parameters, and the active and passive parties share the final training results of the overall model. The process of the SecureBoost algorithm [8] on a single decision tree is shown in Algorithm 1.

Algorithm 1 SecureBoost

Input:

I, Sample space of the current node ;

$\{G^i ⊡ H^i\}_{i=1}^{m}$, Aggregated encrypted gradient statistics obtained from m participating parties

Output:

Partitioning the current sample space based on
threshold of selected sample features

1 Initialize parameters

2 **Active party performs :**

3      $g \leftarrow \sum_{i \in I} g_i, h \leftarrow \sum_{i \in r} h_i$

4   //Iterating over all participating parties

5   **for** $i = 0 \rightarrow m$ **do**

6   1 //Iterating over all features of participating party i

7   2 **for** $k = 0 \rightarrow d_i$ **do**

8   3 $g_i \leftarrow 0, h_i \leftarrow 0$

9   4 //Iterating over all threshold values of k

10   5 **for** $v = 0 \rightarrow l_{k_i}$ **do**

11   6   Obtaining the decryption values $D\left(G_{k,v}^i\right)$ and $D\left(H_{k,v}^i\right)$

12   7 $g_l \leftarrow g_l + D\left(G_{k,v}^i\right), h_l \leftarrow h_l + D\left(H_{k,v}^i\right)$

13   8 $g_r \leftarrow g - g_l, h_r \leftarrow h - h_l$

14   9 $score \leftarrow \mathbf{max}\left(score, \dfrac{g_l^2}{h_l + \lambda} + \dfrac{g_r^2}{h_r + \lambda} - \dfrac{g^2}{h + \lambda}\right)$

15   10 **end for**

16   11 **end for**

17   **end for**

18   When the maximum score is obtained, return k_{opt} and v_{opt} to the corresponding **Passive party** i

19   **Passive party** i **performs:**

20   Determine the threshold for the selected feature based on k_{opt} and v_{opt}, partition the current sample space

21   Record the threshold of the found feature in the lookup table and return the record id and I_L to the **Active party**

22   **Active party performs :**

23   Split the current node according to I_L and associate the current node with [participant id, record id]

24 Return

Specifically, in this paper, the SecureBoost algorithm is used to perform sample alignment on green credit data of listed companies from different datasets with the same sample ID setting. The SecureBoost algorithm is also adopted to expand sample features. SHAP values are then used to analyze the feature results output by the SecureBoost algorithm and select corresponding indicators to generate the expanded esgcredit dataset.

3.2 TCA and BDA Algorithms

The Transfer Component Analysis (TCA)[6] algorithm, as a classic feature-based transfer algorithm, aims to learn the transferable features between the target domain and the source domain by minimizing domain differences. TCA algorithm mainly focuses on calculating MMD [12] (maximum mean discrepancy distance) and generates high-dimensional matrices, and then minimizes MMD through corresponding constraints. The maximum mean discrepancy is defined as follows:

Assuming the given source domain Xs and target domain Xt, and the corresponding source outputs Ys and target outputs Yt, it holds $X_s \cap X_t \neq \varnothing$ and $Y_s = Y_t$, Regarding Xs and Xt, we have:

$$MMD(X_s, X_t) = tr(KL) \tag{1}$$

where K is a composite kernel matrix $K = \begin{bmatrix} K_{s,s} & K_{s,t} \\ K_{s,t}^T & K_{t,t} \end{bmatrix} \in \mathbb{R}^{(n_s+n_t) \times (n_s+n_t)}$, consisting of the source domain kernel matrix $K_{s,s}$, target domain kernel matrix $K_{t,t}$, and cross-domain kernel matrix $K_{s,t}$. In the equation, L is also a matrix, and its elements l_{ij} are defined as follows:

$$l_{ij} = \begin{cases} \frac{1}{n_s^2} & x_i, x_j \in X_s \\ \frac{1}{n_t^2} & x_i, x_j \in X_t \\ -\frac{1}{n_s n_t} & else \end{cases}$$

The TCA algorithm has the following constraints on the MMD distance:

$$\min_W tr\left(\tilde{K} WW^T \tilde{K} L\right) + \lambda tr\left(W^T W\right)$$
$$s.t. W^T \tilde{K} H\tilde{K}W = I \tag{2}$$

Equation (2) mainly aims to minimize the MMD distance between the mapped source and target domains, while imposing constraints on K to ensure that the transferred data features are within a reasonable range for comparison.

Compared to the TCA algorithm, the Balanced Distribution Adaptation (BDA)[7] algorithm considers the distribution of different categories between the source and target domains and uses category information to balance the distribution between the domains to improve the effectiveness of transfer learning. The core of the BDA algorithm is to set two optimization objectives that consider both the shortest distance between the source and target domains and the optimal distance between different category distributions, thereby improving the effectiveness of feature transfer. The constraint function for different category distributions is defined as follows:

Let $M = [m_{i,j}]$ be the correlation matrix of $n_c \times n_c$, where the matrix element $m_{i,j}$ represents the correlation between the i-th category in the source domain and the j-th category in the target domain, we have:

$$\min_M \sum_{i=1}^{n_s} \sum_{j=1}^{n_t} w_{i,j} \left\| f\left(x_{s,i}; W\right) - f\left(x_{t,j}; W\right) \right\|_2^2 m_{y_{s,i}, yt,j} \tag{3}$$

Let $w_{i,j}$ be the weight between the i-th sample in the source domain and the j-th sample in the target domain, and $m_{y_{s,i}, yt,j}$ represents the correlation between the $y_{s,i}$-th category in the source domain and the $y_{t,j}$-th category in the target domain. By imposing this constraint, the BDA algorithm solves for the optimal distance between category distributions, thereby improving the effectiveness of multi-class transfer learning.

It should be noted that because the BDA algorithm imposes constraints on the optimal distance between category distributions, it often requires a similar number of samples for each classification label. When faced with datasets with large differences in sample sizes between classification labels, the performance of the BDA algorithm may be affected, and it tends to perform better on classifications with more data than those with less data.

3.3 Feature Transfer Learning Based on Improved-TCA

This paper proposes a feature transfer model based on an improved TCA algorithm by integrating the characteristics of both TCA and BDA algorithms.

By considering the difference in the number of classification samples in different labeled samples of real datasets, this paper sets parameters α and β to reflect the difference in the magnitude of data quantity among classification samples. The outputs of TCA and BDA algorithms are conditionally aggregated. Specifically, this paper designs a feature transfer model based on the improved TCA algorithm, which demonstrates superior performance in the case of imbalanced sample distribution in the dataset (see Fig. 2).

α represents the ratio of the data volume between small-sample size classification labels and large-sample size classification labels, while β represents the ratio of the sample label data volume to the data volume of the source domain dataset.

When the classification label belongs to a large sample size classification label, the final prediction result is weighted by αTCA + $(1-\alpha)$BDA, and the weighted result is rounded off. At this time, the result of the BDA algorithm has a greater impact on the transfer prediction results for the classification. When the classification label does not belong to a large sample size classification label, the final prediction result is weighted by $(1-\beta)$TCA + βBDA, and the weighted result is rounded off. At this time, the result of the TCA algorithm has a greater impact on the transfer prediction results for the classification. By judging the sample size of the classification label, we combine the better performance of the BDA algorithm on the large sample classification labels with the TCA algorithm. Under the influence of the parameter α and β, BDA and TCA perform joint transfer training between the source and target domain datasets, and finally obtain the classification label of the target domain dataset.

Based on the judgment of the distribution of classification label sample data volume, we combined the TCA and BDA algorithms, improved the TCA algorithm by adopting the idea of optimal distance between sample distributions in the BDA algorithm, and constructed a feature transfer model based on the improved TCA algorithm. The proposed model effectively improves the transfer prediction of classification labels with larger data volumes under uneven distribution of sample label data volumes, and also improves the transfer training results for classification labels with smaller data volumes, effectively solving the problem of poor transfer results of TCA algorithm on classification labels with larger data volumes and distorted transfer results of BDA algorithm on classification labels with smaller data volumes.

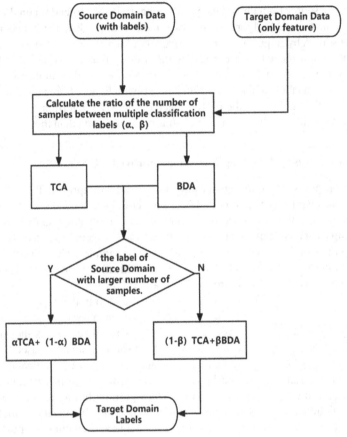

Fig. 2. Design of Feature Transfer Model based on the improved TCA algorithm

4   Experimental Results

This section will introduce the experimental environment, dataset, and corresponding experimental results of this study. The experiments in this study were divided into two sections: vertical federated learning experiments and feature-based transfer learning experiments. XGBoost algorithm and random forest algorithm were used as benchmark experimental algorithms in this study, while SecureBoost algorithm was used as the algorithm for vertical federated learning experiments, and TCA, BDA, and Improved-TCA were used as the algorithms for feature-based transfer learning experiments. Specifically, ACC, Precision, and Recall values were used to evaluate the experimental results of different algorithms on the dataset. For the experimental results on different classification labels, Precision and Recall were used for evaluation.

4.1 Experimental Environment

The experimental environment of this study was implemented in a Linux environment running under Ubuntu 20.04 simulated by VMware. The FATE 1.6.0 framework was installed using Docker to simulate client and server-side training using the SecureBoost algorithm in vertical federated learning. The algorithms, TCA, BDA, and Improved-TCA were compiled using Anaconda in Jupyter Notebook based on Python3.8. Furthermore, the codes for benchmark experiments such as XGBoost and random forest were also compiled, and the feature-based transfer learning was trained.

4.2 Dataset

The experimental datasets selected in this study were primarily composed of the wind and huazheng datasets, which were derived from the wind green finance database, Huazhong ESG rating database, and Guotai An database.

The wind green finance database covers most of the A-share listed companies and divides the score into scores reflecting the long-term fundamental impact of environmental, social, and governance aspects and scores reflecting the controversy events. The overall green credit score for listed companies is then classified into seven levels ranging from AAA to CCC. In this study, the wind2021 and wind2022 datasets, which include 1,893 listed companies, were used.

The Huazhong ESG rating database discloses the ESG rating levels and environmental, social, and governance rating levels of A-share listed companies for the past two years, and divides the company rating level into nine levels ranging from AAA to C. In this study, the huazheng2021 and huazheng2022 datasets, which include 4,937 listed companies, were used.

Prior to conducting the experiments, data preprocessing was performed on the datasets, and the corresponding levels were mapped. Sample IDs with identical features were selected from each dataset as the composition of the dataset for experimentation. The distribution of the final experimental datasets is shown in the Table 1.

Among them, classification 1 and classification 2 belong to the large sample classification, while the rest are mostly small sample classifications.

4.3 Vertical Federated Learning

For different datasets of the same year, we employed SecureBoost algorithm in vertical federated learning to predict five-classification labels, and compared the experimental results with XGBoost and Random Forest as baseline methods. The final results are shown in Table 2.

Compared to XGBoost and Random Forest, SecureBoost algorithm achieves better ACC and Precision results. In comparison to the results of XGboost algorithm on the wind dataset, SecureBoost algorithm improves the ACC by about 7% and Precision by about 8%. The SecureBoost algorithm achieves privacy protection through homomorphic encryption and trains the indicators of the wind dataset and the huazheng dataset together, resulting in better performance than training on a single dataset alone. Compared to the

Table 1. Label mapping and distribution of grade indicators in wind and huazheng datasets.

Dataset	level	labels	meaning	Data-2021	Data-2022
wind	AAA	4	excellent	61	53
	AA				
	A	3	good	199	223
	BBB	2		489	507
	BB	1		541	553
	B	0	qualified	98	52
	CCC				
huazheng	AAA	4	excellent	106	23
	AA				
	A				
	BBB				
	BB	3	good	375	165
	B	2		465	448
	CCC	1		264	591
	CC	0	qualified	178	161
	C				

Table 2. Five-classification prediction results of XGBoost, Random Forest and SecureBoost on datasets.

Model		2021		2022	
		wind	huazheng	wind	huazheng
XGBoost	ACC	0.7636	0.8103	0.7721	0.8349
	Precision	0.7621	0.8043	0.7697	0.8201
	Recall	0.6437	0.7686	0.6814	0.7739
Random Foreset	ACC	0.7461	0.7511	0.7547	0.7598
	Precision	0.7310	0.8431	0.7402	0.8317
	Recall	0.6221	0.6121	0.6719	0.6309
Secure Boost	ACC	**0.8341**		**0.8427**	
	Precision	**0.8382**		**0.8453**	
	Recall	**0.6630**		**0.6929**	

traditional XGBoost and random forest algorithms, the results of SecureBoost are more stable, with significant improvements in ACC and Precision.

Regarding the results output by the SecureBoost algorithm, we conducted SHAP value analysis. Taking the output results of SecureBoost after conducting vertical federated learning on the wind2022 dataset and the huazheng2022 dataset as an example, the SHAP value explanations for the top eight features ranked by importance on the model output are shown in Fig. 3.

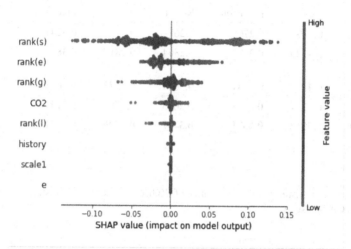

Fig. 3. The distribution of SHAP values for the indicators in the esgcredit2022 dataset.

The scattered points in the figure represent the influence of each feature on the model's final output. The color of the point varies from blue to red, indicating a low to high SHAP value for the corresponding feature. The feature indicators in the figure are arranged from top to bottom in descending order of importance level, with "rank(s)" being the most important and "e" being the least important. The figure uses 0 as the boundary to judge the positive or negative impact of an indicator on the model output.

Analyzing "rank(s)", "rank(e)", and "rank(g)" using their respective SHAP values, we found that as the SHAP value of a feature increases, its impact on the model output leans more toward positive values, indicating a larger positive influence on the model output. On the other hand, when analyzing "CO_2", we found that as the SHAP value increases, the indicator data tends to be more negative, indicating a certain negative impact on the model's final output.

Based on the analysis of the SHAP values, we used the top eight indicators ranked by importance in the SecureBoost training in Fig. 3 to construct the "esgcredit" dataset as one of the experimental datasets for feature migration in the future.

4.4 Feature-Based Transfer Learning

For datasets of different years, we used the 2021 dataset as the source domain and the 2022 dataset as the target domain. We conducted experiments on three datasets using

XGBoost, TCA, BDA, and Improved-TCA algorithms proposed in this paper. The overall experimental results are shown in Table 3 and Table 4.

Table 3. Feature transfer prediction results of XGBoost, Random Forest,TCA, BDA, and Improved-TCA on datasets

Model	Wind			Huazheng		
	ACC	Precision	Recall	ACC	Precision	Recall
XGBoost	0.387	0.2708	0.261	0.2714	0.3241	0.2531
Random Forest	0.4123	0.3587	0.2804	0.2817	0.3841	0.3139
TCA	0.4373	0.3741	0.3614	0.3573	0.3653	0.4221
BDA	0.5144	0.2098	0.2643	0.2798	0.3141	0.3127
Improved TCA	**0.5512**	**0.4671**	**0.3447**	**0.4596**	**0.5831**	**0.3921**

Table 4. Feature transfer prediction results of XGBoost, Random Forest, TCA, BDA, and Improved-TCA on esgcredit datasets

Model	esgcredit		
	ACC	Precision	Recall
XGBoost	0.487	0.4618	0.5007
Random Forest	0.5095	0.3647	0.4048
TCA	0.5245	0.4171	0.4508
BDA	0.5764	0.3401	0.3126
Improved TCA	**0.6318**	**0.6238**	**0.5098**

From the experimental results, the Improved-TCA-based feature transfer model proposed in this paper has overall better performance in ACC, Precision, and Recall compared to other algorithms. Comparing with the performance of the TCA algorithm on the esgcredit dataset, the ACC of the proposed Improved-TCA-based feature transfer model is improved by 10.73%, Precision by 20.67%, and Recall by 5.90%.

Comparing results from different datasets, the esgcredit dataset generated by vertically federated learning through the SecureBoost algorithm in this paper performs better in feature transfer than the wind and huazheng datasets. Taking the performance of the Improved-TCA algorithm on three datasets as an example, the overall label values of the esgcredit dataset are about 10% higher in ACC, Precision, and Recall than those of wind and huazheng. Therefore, the experimental results prove that datasets processed by vertically federated learning perform better in transfer training, and vertically federated learning can be combined with feature transfer algorithms to output better results.

Compared with traditional machine learning algorithms XGBoost and Random Forest, the feature-based transfer learning algorithms TCA, BDA, and Improved-TCA-based feature transfer models perform better in ACC. The Improved-TCA-based feature transfer model and the TCA algorithm perform better in Recall and Precision than the BDA algorithm. The overall results of Precision and Recall in separate classification for TCA and Improved-TCA-based feature transfer models are shown in Fig. 4.

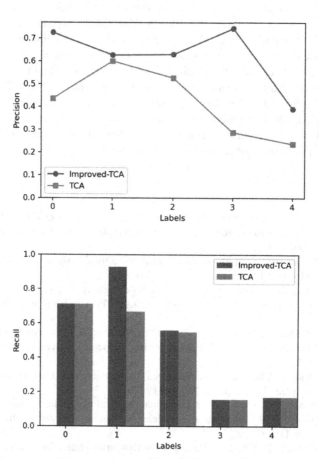

Fig. 4. The Precision and Recall of Improved-TCA and TCA on different classification labels

Overall, the experimental results show that the feature transfer model based on SecureBoost and Improved-TCA proposed in this paper performs better than traditional machine learning algorithms used for feature transfer in the absence of vertical federated learning. Comparing the transfer results of XGBoost on the wind dataset with Improved-TCA on the esgcredit dataset, the ACC value of Improved-TCA is improved by 24.48%, the Precision value is improved by 35.30%, and the Recall value is improved by 24.88%.

The experiment confirms that the green credit risk control model based on Secure-Boost and Improved-TCA proposed in this paper has strong practical significance in the field of green credit rating. It also further validates that when dealing with imbalanced data distribution, using the feature transfer model based on Improved-TCA proposed in this paper is superior to TCA and BDA algorithms in terms of the experimental results.

5 Conclusion

In this paper, we propose a green credit risk control model based on SecureBoost and Improved-TCA, which optimizes the five-class classification problem of green credit rating for listed companies by combining vertical federated learning and feature transfer. We improve the TCA algorithm by setting hyperparameters α and β to adjust the data label sample ratio, weighting the results of the TCA and BDA algorithms in different classifications to obtain the feature transfer model based on Improved-TCA. Through experiments on three datasets, we verify that the method of combining vertical federated learning with feature-based transfer learning performs better in feature transfer. Subsequently, we will conduct experiments on other publicly available datasets with imbalanced classification label distribution.

References

1. Zhang, W., Zhang, Y.X.: Review of measuring and evaluating green credit and its influence literature. Cooperation Econ. Technol. **697**(02), 56–58 (2023). https://doi.org/10.13665/j.cnki.hzjjykj.2023.02.045
2. Zhao, Q.X.: The impact of green credit on the operational performance and risk-taking of commercial banks. China Price **408**(04), 77–80 (2023)
3. Shu, L.M., Liao, J.H., Xie, Z.: Green credit policy and enterprise green innovation: empirical evidence based on the perspective of green industry. Financ. Econ. Res. **38**(02), 144–160 (2023)
4. Long, M., Wang, J., Ding, G., et al.: Transfer feature learning with joint distribution adaptation. In: Proceedings of the IEEE Conference on Computer Vision and Pattern Recognition, pp. 5147–5154 (2014)
5. Yang, Q., Liu, Y., Chen, T., et al.: Federated machine learning: concept and applications. ACM Trans. Intell. Syst. Technol. **10**(2), 1–19 (2019)
6. Pan, S.J., Tsang, I.W., Kwok, J.T., et al.: Domain adaptation via transfer component analysis. In: International Joint Conference on Artificial Intelligence. Morgan Kaufmann Publishers Inc., pp. 1186–1191 (2009)
7. Wang, J., Chen, Y., Hao, S., et al.: Balanced distribution adaptation for transfer learning. IEEE Int. Conf. Data Min., 1125–1130. IEEE (2017)
8. Cheng, K., Fan, T., Jin, Y., et al.: SecureBoost: a lossless federated learning framework. IEEE Intell. Syst., (99), 1–1 (2021)
9. Merrick, L., Taly, A.: The explanation game: explaining machine learning models using shapley values. arXiv preprint arXiv:1903.00031 (2019)
10. Chen, T., Guestrin, C.: XGBoost: a scalable tree boosting system. Proc. ACM SIGKDD Int. Conf. Knowl. Discov. Data Min. **13**(1), 1–15 (2016)
11. Sun, Q., Amin, M., Yan, B., et al.: Transfer learning for bilingual content classification. In: Proceedings of the ACM SIGKDD International Conference on Knowledge Discovery and Data Mining, 1531–1540

12. Scannapieco, M., Figotin, I., Bertino, E., et al.: Privacy preserving schema and data matching. In: ACM Sigmod International Conference on Management of Data. ACM (2007)
13. Gretton, A., Borgwardt, K.M., Rasch, M., et al.: A kernel two-sample test. J. Mach. Learn. Res. **13**, 723–773 (2012)
14. Dai, L.Y., Li, C., Mao, M.H.: Heterogeneous efficiency incentive of green credit on commercial banks: an empirical study based on meta-frontier DEA framework. J. Hainan University (Social Science Edition), 1–11 (2023). https://doi.org/10.15886/j.cnki.hnus.202209.0210
15. Yang, Q.: Federated Learning. Electronic Industry Press, Beijing (2020). (in Chinese)
16. Lundberg, S., Lee, S.I.: A unified approach to interpreting model predictions. Adv. Neural. Inf. Process. Syst. **30**, 4765–4774 (2017)
17. Wang, J., Chen, Y., Feng, W., et al.: Transfer learning with dynamic distribution adaptation. ACM Trans. Intell. Syst. Technol. **11**(1), 1–25 (2020)

Unsupervised Concept Drift Detection Based on Stacked Autoencoder and Page-Hinckley Test

Shu Zhan[1], Yang Li[2,3], Chunyan Liu[1], and Yunlong Zhao[1]([✉])

[1] College of Computer Science and Technology, Nanjing University of Aeronautics and Astronautics, Nanjing, China
zhaoyunlong@nuaa.edu.cn
[2] Unmanned Aerial Vehicles Research Institute, Nanjing University of Aeronautics and Astronautics, Nanjing, China
[3] Key Laboratory of Advanced Technology for Small and Medium-Sized UAV, Ministry of Industry and Information Technology, Nanjing, China

Abstract. Data streams are often subject to concept drift, which can gradually reduce the reliability of learning models over time in data stream mining. To maintain model accuracy and enhance its robustness, it is crucial to detect concept drift and update the learning model accordingly. The majority of drift detection methods rely on the assumption that true labels are immediately available, which is challenging to implement in real-world scenarios. Therefore, it is more practicable to detect concept drift in an unsupervised manner. This paper proposes an unsupervised Drift Detection method based on Stacked Autoencoder and Page-Hinckley test (DD-SAPH). DD-SAPH employs the stacked autoencoder as a medium to represent the distribution of historical data, which extracts hidden features from the reference window. To measure the difference between distributions of historical data and new data, the reconstruction error of the stacked autoencoder on the current window is employed. The Page-Hinckley test dynamically calculates thresholds to warn and alarm concept drift. Experimental results indicate that DD-SAPH outperforms the compared unsupervised algorithms when addressing concept drift on both synthetic and real datasets.

Keywords: Data stream mining · Unsupervised learning · Concept drift detection · Stacked autoencoder · Page-Hinckley test

1 Introduction

Mining and utilizing meaningful information from data streams has become one of the most significant research areas. Frequently, it is assumed that the process of creating such a data stream is stationary [21]. However, in real-world scenarios, the data stream evolves over time, which means that the knowledge pattern that data adheres to may change. The term for this phenomenon is concept drift [24]. Concept drift makes the learning model created by using historical data

© The Author(s), under exclusive license to Springer Nature Singapore Pte Ltd. 2024
H. Jin et al. (Eds.): GPC 2023, LNCS 14503, pp. 192–208, 2024.
https://doi.org/10.1007/978-981-99-9893-7_15

ineffective on fresh data [20]. Failure to update the model promptly when drifts are present results in a constant increase in the number of errors [25]. Rather than assuming the data is in a stationary environment, a successful learning model should continually detect concept drift in the datastream and be updated or retrained to adapt to changing data distributions [12].

Most existing detection methods involve continuously monitoring the performance of the downstream base model and delivering drift alerts when the model's performance degrades substantially. Although highly accurate, these methods rely on sample labels to calculate the model's performance, assuming that labels are readily available. However, in real-world scenarios, labels may arrive late or not at all due to the high cost of tagging [30]. In this context, an increasing number of researchers are focusing on unsupervised concept drift detection approaches that do not require labeled information [9]. Traditional unsupervised conceptual drift detection algorithms require dividing the data stream into multiple data chunks and using the summarized information (e.g., frequency histograms) in each chunk as a reference distribution, and using distributional discrepancy measures, such as the Hellinger distance and the Kullback-Leibler scatter, to measure the difference between the distributions of the two multivariate data chunks. Finally, the idea of hypothesis testing is used to determine whether drift has occurred based on the difference measure. In this paper, we will abandon this traditional statistical idea and instead use deep learning techniques to train models to learn data distributions. A stack autoencoder is introduced to utilize its powerful feature learning capability to train the model to self-learn to the distributional features about the training data in this paper. And a balance between robustness and sensitivity to the perception of data distribution differences is achieved by introducing a dynamic threshold setting approach.

This study introduces a novel unsupervised concept drift detection algorithm named DD-SAPH. First, DD-SAPH generates a reference window from a subset of instances representing historical data and divides it into numerous data blocks containing distribution information. Using the autoencoder's data correlation, each data block is used to guide the training of a stacked autoencoder (SAE). The trained SAE extracts hidden features describing the data distribution in the reference window that are closely related to the initially known concept. Following this, the current data instances to arrive comprise the current window. The reconstruction error resulting from the attempt to match the distribution of the data in the current window is used as a measure of the difference between the distributions of the current window and the reference window. Finally, the Page-Hinckley test (PH test) [32] dynamically calculates the drift threshold and flags drift when the difference in distributions exceeds the threshold.

The subsequent organization of the paper is as follows: Sect. 2 reviews the current state of research related to concept drift detection. Section 3 introduces the unsupervised concept drift detection DD-SAPH. Section 4 details the performance evaluation results of DD-SAPH on synthetic and real datasets. Section 5 concludes the paper and discusses the future direction.

To summarise, the contributions of this paper are the following:

- DD-SAPH: a novel method for unsupervised concept drift detection based on stacked autoencoder and Page-Hinckley test
- DD-SAPH proposes for the first time the use of a stacked self-encoder to automatically learn the data distribution between windows, and combines it with dynamic threshold settings to achieve a balance between robustness and sensitivity to the perception of distribution changes.
- A set of experiments used to validate the proposed method. These experiments include comparisons with supervised detection algorithms and unsupervised detection algorithms.

2   Related Work

Concept drift detection algorithms can be categorized into three groups based on their dependence on labeled data for detection: supervised, semi-supervised, and unsupervised.

2.1   Supervised Detection Algorithms

Supervised detection algorithms typically rely on labeled data to continuously track changes in performance metrics, such as error rate, recall, F-score, etc., to detect the occurrence of concept drift.

The Drift Detection Method (DDM) proposed by Gama et al. [14] is one of the most well-known methods and is utilized as a baseline by numerous algorithms. DDM checks the classifier's error rate throughout the classification task. Increasing error rates indicate the presence of concept drift. Baena-García et al. proposed the Early Drift Detection Method (EDDM) to overcome the DDM's insensitivity to gradual changes [3]. EDDM is an extension of DDM that is superior to DDM in the face of gradual drift. EDDM is not concerned with the error rate of the classifier but rather the distance between misclassified instances. The Hoeffding Drift Detection Method (HDDM) was proposed by Frias-Blanco et al. [13] in 2014. HDDM monitors the performance metrics of the classifier using probabilistic inequalities, offering a theoretical guarantee that the false negative rate and false positive rate will be maintained within a specific bound. DDM, EDDM, and HDDM are all based on Statistical Process Control (SPC), a standard statistical method for monitoring product quality in a continuous production process.

ADWIN [4], DoD [31], and KSWIN [28] partition the set of performance metrics at various timestamps into many windows and apply different criteria to evaluate if drift has occurred by comparing the differences between the windows. These approaches exhibit robustness against noise but necessitate a certain amount of memory to store information in the window.

2.2   Semi-supervised Detection Algorithms

Semi-supervised detection algorithms, which are partially label-dependent, utilize limited label information to accept concept drift-containing data streams.

Ahmadi and Beigy present a semi-supervised integrated learning system that can accommodate concept drift for data streams [2]. The approach uses the majority vote of the classifiers to label unlabeled instances, and the labeled instances are utilized to update each single classifier. In the method proposed by Hosseini, Gholipour, and Beigy [18], a pool of classifiers is maintained, with each representing a distinct notion. The method labels a portion of these instances and updates the classifiers in the pool using the labeled partial batch to identify which ones are most similar to the underlying notion of the batch. A new classifier will be trained and added to the pool if not. SAND [16] uses a classifier to identify concept drift by monitoring the confidence in classification of the input unlabeled data. The algorithm preserves the integration of classifiers that are each trained on a dynamically determined block. The similar cohesion measurement between unclassified outliers is employed to detect concept drift that arises from the emergence of new classes.

2.3 Unsupervised Detection Algorithms

Unsupervised detection algorithms work by keeping track of changes in the way data features are spread out, and they can be done without any label information. Kifer et al. first developed in 2004 the most generic method [19] for analyzing differences between distributions of two data windows in a data streaming scenario, hence guiding the development of distribution distance functions. In real-world scenarios where the data do not adhere to a certain distribution, non-parametric hypothesis testing approaches are chosen. Ross, Tasoulis, and Adams [29] generate a streaming change-point model (CPM) that employs a number of rank-based nonparametric hypothesis tests for univariate data streams with arbitrary continuous distributions. But this method is impractical and only applicable to one-dimensional data streams. When monitoring the distribution of multivariate data, it is common practice to segment the data stream into multiple blocks and present the distribution of each block using summary information such as frequency histograms. Several techniques, including Hellinger distance, KL-divergence, and others, have been employed to quantify the dissimilarity between the distributions of two multivariate data blocks. HDDDM [11] uses histograms to approximate the distribution within every data block and the Hellinger distance to detect potential changes in distribution between data blocks. The method [10] proposed by Dasu et al. is the first information-theoretic system for change detection with a strong statistical basis. To assess the difference between two distributions, the proposed approach employs the KL-divergence, followed by the application of the percentile bootstrapping technique [32] to determine the statistical significance of the differences. To partition the space, the approach uses a kdq-tree. D3 [15] employs a discriminant classifier as a concept drift detector and labels the data in the window according to their arrival order. Using the labeled data to train the discriminant classifier, the AUC will be utilized to detect if the distribution between the old and new samples has changed.

In addition, the technique based on Principal Component Analysis (PCA) aids in enhancing the algorithm's performance with high-dimensional data [27]. PCA-CD projects a high-dimensional data stream into a low-dimensional format via PCA, making density estimates and change-score computations easier. The change-scores on each dimension are then computed using the KL-divergence, and drift is indicated when the highest change-score exceeds a threshold set using the PH test. This method decreases the number of features that need to be observed and proves that monitoring a smaller feature space may detect drift in the original features.

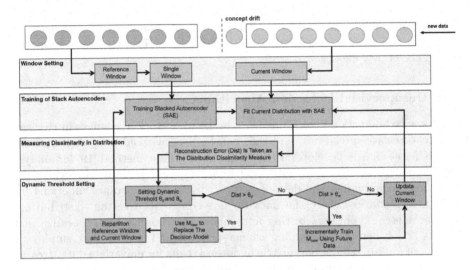

Fig. 1. Overall architecture of DD-SAPH

3   Method

In this paper, we present DD-SAPH, an unsupervised concept drift detection method, which is illustrated in Fig. 1. The architecture consists of four separate components, namely window setting, training of SAE, measuring dissimilarity in distributions, and dynamic threshold setting. The pseudo-code implementing DD-SAPH is shown in Algorithm 1 to give additional, more specific details.

3.1   Window Setting

Setting windows is a common way to store data when it comes to collecting and organizing information. Our method includes three types of windows: the reference window S_{ref}, the current window S_{cur}, and the distribution representation window S_{single}(Lines 3–4 and 12–15 in Algorithm 1). S_{ref} consists of

---

Algorithm 1. Concept drift detection algorithm DD-SAPH

Require: streaming data S, window sizes w_1, w_2, sliding steps L, λ_d, λ_w, δ for PH test

Ensure: Report a change at time t when detecting a change

1: Initialize $Train$, S_{ref}, S_{single}, S_{cur} to NULL
2: Initialize $t_c = m = M = 0$
3: Set $posR = t_c$, $posS = t_c$
4: Set $S_{ref} = \{X_{posR+1}, \cdots, X_{posR+w_1}\}$
5: **while** $posS + w_2 \leq posR + w_1$ **do**
6:     $S_{single} = \{X_{posS+1}, \cdots, X_{posS+w_2}\}$
7:     $A = Flatten(S_{single})$
8:     $Train = Train \bigcup A$
9:     $posS = posS + L$
10: **end while**
11: Using $Train$ to training SAE
12: **while** a new sample X_t arrives in the stream **do**
13:     **while** $len(S_{cur}) < w_2$ **do**
14:         $S_{cur} = S_{cur} \bigcup X_t$
15:     **end while**
16:     $Dis = Dist\left(Flatten\left(S_{cur}\right)\right)$ using Eq. (3)
17:     **if** $threhold\ making\ by\ PH\ test = drift$ **then**
18:         Report a change at time t
19:         replace classifier with M_{new}
20:         $t_c = t, posR = posS = t - \lfloor \frac{w_2}{2} \rfloor$
21:         clear S_{cur} and GOTO step 3
22:     **else if** $threhold\ making\ by\ PH\ test = warning$ **then**
23:         Train new classifier M_{new} with all data comes after
24:     **else**
25:         Remove the first L samples in S_{cur} and reset M_{new}
26:     **end if**
27: **end while**

---

the first w_1 data arrivals following the occurrence of the last drift and represents the initial data distribution. To account for recurring concept drift, the reference window needs to be updated correspondingly. The most recent w_2 instances are included in S_{cur}, which is more reflective of the current data distribution. S_{single} is the smallest data batch utilized to characterize the data distribution, which is used for training SAE. It has the same window capacity as the current window S_{cur}. The window's capacity invariably influences the method's performance. Typically, the window size is determined by the application problems.

3.2 Training of SAE

To estimate the data distribution indirectly, SAE is trained using the data in the reference window. A well-learned SAE can extract features that effectively capture the underlying distribution information of the batch data. The training process of SAE is shown in the first component of Fig. 2.

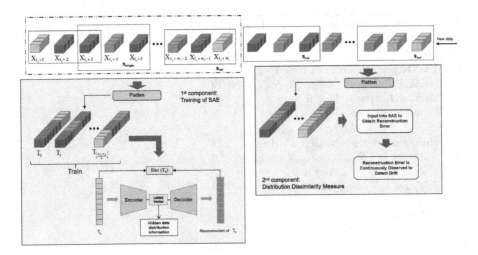

Fig. 2. The process of training stacked autoencoder (SAE) and measuring dissimilarity in distributions

An important aspect of enabling SAE to extract distribution information is determining how to use the data in the reference window to produce a suitable training set. Let the moment of the last concept drift be t_c. The first w_1 samples after the t_c are used to populate S_{ref}, which is $S_{ref} = \{X_{tc+1}, X_{tc+2}, \cdots, X_{tc+w_1}\}$, where $X_i \in R_p$, p is the dimension of the original data. After getting the reference window, the training set $Train$ of SAE is produced as follows(Line 5−10 in Algorithm 1):

First, initialize S_{single} to be the first w_2 data from S_{ref}, where $w_2 \ll w_1$. Then, S_{single} slides gradually along the direction of data arrival until $a < L$. L is the number of the sliding steps and a represents the number of data instances in S_{ref} that have yet to enter S_{single}. To be specific, $S_{single} = \{X_{tc+i}, X_{tc+i+1}, \cdots, X_{tc+i+w_2-1}\}$, where $i = 1, L+1, 2*L+1, \cdots, \lfloor\frac{w_1-w_2}{L}\rfloor * L + 1$. Finally, the various data batches in S_{single} corresponding to each value of i will be added to the training set $Train$ as a one-dimensional vector, and the final training set will be $\lfloor\frac{w_1-w_2}{L}\rfloor + 1$ in size. The dimension of each training data in the $Train$ is $p*w_2$ in the end, which means $Train = \left\{T_0, T_1, \cdots, T_{\lfloor\frac{w_1-w_2}{L}\rfloor}\right\}$, where $T_i \in R_{p*w_2}$.

After the production of the training set, the SAE training process is initiated(Line 11 in Algorithm 1). Encoder and decoder are the two components of a SAE, which has $L+1$ layers. The input and output layers are both D in size. The j-th neuron of the l-th layer has the following equation:

$$y_j^{(l)}(T_n) = \sigma\left(\sum_{i=1}^{N_{l-1}} w_{ij}^{(l)} y_j^{(l-1)}(T_n) + b_j^{(l)}\right) where \quad l = 1, 2, 3, \cdots, L \quad (1)$$

T_n ($T_n \in R_{p*w_2}$) is the input vector, which is a data element in $Train$. T_n^j is the j-th component of the input vector, and $y_j^{(0)} = T_n^j$. N_{l-1} is the number of neurons in the $l-1$-th layer. $w_{ij}^{(l)}$ is the weight between the $l-1$-th layer and the l-th layer. $b_j^{(l)}$ is the bias of the l-th layer. $\sigma(z)$ is the activation function. We choose $LeakyReLU$ as the activation function, which not only has the advantages of $ReLU$ function to alleviate the gradient disappearance and converge faster, but also can solve the neuron death problem of $ReLU$. Its mathematical expression is as follows, where $leak$ is a small constant:

$$\sigma(z) = max(0, z) + leak * min(0, z) \tag{2}$$

Training SAE has as its primary optimization goal making the output of the network infinitely close to the input and obtaining a better representation, i.e., minimizing the loss function $Dist(T_n)$, also known as reconstruction error:

$$Dist(T_n) = Dist\left(T_n, y_j^{(L)}(T_n)\right) = \left(\sum_{j=1}^{D} (T_n^j) - y_j^{(L)}(T_n)\right)^2 \tag{3}$$

To minimize the loss function, the back propagation algorithm with gradient descent optimization is used to guide the training of SAE. The weight $w_{ij}^{(l)}$ and the bias $b_j^{(l)}$ are updated according to the following equation:

$$w_{ij}^{(l)} := w_{ij}^{(l)} - \eta \frac{1}{B} \sum_{m=t}^{t+B-1} \frac{\partial Dist(T_m)}{\partial w_{ij}^{(l)}}$$

$$b_j^{(l)} := b_j^{(l)} - \eta \frac{1}{B} \sum_{m=t}^{t+B-1} \frac{\partial Dist(T_m)}{\partial b_j^{(l)}} \tag{4}$$

where η is the learning rate, B is the batch size. The current training data batch is $\{T_t, T_{t+1}, \cdots, T_{t+B-1}\}$.

3.3 Measuring Dissimilarity in Distributions

As new data comes in, the difference between the current data distribution and the historical data distribution is measured using the reconstruction error, as depicted in the second component of Fig. 2.

Enable the current window S_{cur}; S_{cur} slides continuously in the direction of the incoming data stream with steps L, always holding the most recent incoming w_2 data. Using the same approach as in Sect. 3.2 for S_{single}, the data from the current window is transformed into a one-dimensional vector and fed into SAE, after which the reconstruction error (Eq. 3) is calculated. When the data in the current window and the reference window share the same distribution, the neural unit in the hidden layer of the encoder receives accurate distribution information. Subsequently, the decoder then decodes this concealed distribution information and restores it such that it is virtually identical to the original data. As a result, the reconstruction error will vary up and down around a very modest number dif (close to zero). Otherwise, the reconstruction error will significantly deviate from dif. Observing the reconstruction error for this deviation enables the detection of concept drift. A present threshold θ is set to report a change when the current reconstruction error becomes greater than the threshold.

3.4   Dynamic Threshold Setting

As indicated in Sect. 3.3, A change is reported when the current reconstruction error significantly deviates for a reasonable period of time from the history of the reconstruction error values. The occurrence of deviation in the computed difference in distribution (the reconstruction error) is the focus of both our work and that of Qahtan et al. [27] Therefore, we adopt the PH test as used by Qahtan et al. to establish a dynamic threshold for real-time monitoring of concept drift.

PH test defines p_i as the realization of the observed random variable p at moment i [32]. $\overline{p_t}$ represents the empirical average of the p_i at the t moment. m_t is a cumulative variable to store the accumulation of the difference between p_i and $\overline{p_t}$ at the moment t:

$$\overline{p_t} = \frac{1}{t} \sum_{i=1}^{t} p_i \tag{5}$$

$$m_t = \sum_{i=1}^{t} (p_i - \overline{p_i} + \delta) \tag{6}$$

δ is a non-negative real number close to 0, representing the maximum value of the allowed variation. PH test reports the change by continually observing the difference between M_t and m_t, denoted by PH_t. When PH_t is greater than the threshold θ, PH test will report a change, where $M_t = \max\{m_1, m_2, \cdots, m_t\}$

The magnitude of θ defines the trade-off between the model's sensitivity and robustness. A higher θ value makes the model more robust and reduces the incidence of false alarms, but it may raise the likelihood that the model may fail to detect the change point. A smaller θ value makes the model more sensitive and able to detect changes with greater ease, but may result in false positives. The effect of θ should not be disregarded. The method as used by Qahtan et al. [27] is adopted to establish a dynamic threshold. Let $\theta_t = \lambda * \overline{p_t}$, λ is the hyperparameter, called θ-factor. The θ_t at each moment can be self-adjusted based on historical observations.

Lines 17–27 in Algorithm 1 show how to define and utilizing the warning threshold $\theta_w = \lambda_w * \overline{p_t}$ and the drift threshold $\theta_d = \lambda_d * \overline{p_t}$, respectively. Reaching the warning threshold means that the data distribution is predicted to change and a new model M_{new} needs to be trained using the subsequent increment of data samples and maintained. Reaching the drift threshold indicates that a drift has occurred, and the old model is replaced with M_{new}.

4   Experiment

To verify the effectiveness of our method, we will design three sets of experiments to test the following hypotheses respectively(Sections 4.1 to 4.3): H1: The reconstruction error on the current window can accurately estimate the difference in distribution compared to the reference window. H2: DD-SAPH is able to detect changes in data streams having multiple change points with high precision, low latency, and low false positives. H3: DD-SAPH can increase the classification performance of the base classifier more effectively than other unsupervised detection methods when performing a classification.

The architecture of SAE is designed with 7 layers, including 3 layers each for the encoder and decoder, and 1 layer for the hidden variable. In details, the number of neurons in each layer is $P * w_2$ (P is the original data dimension), 512, 64, 10, 64, 512, $P * w_2$, and $leak = 0.01$. For the training of SAE, the batch size of training is set to 64, and the SGD optimizer is used with a learning rate of 0.005. Consistent with the description in Sect. 3.2, the mean square error is used as the objective function.

4.1   Validity of Reconstruction Error for Measuring Dissimilarity in Distributions

Datasets. Artificial datasets make it possible to see if and where the data distribution has changed, thus judging whether the reconstruction error accurately reflects differences in data distributions. Therefore, eight artificial concept drift datasets (including four abrupt and four gradual concept drift datasets) were selected to validate H1 (shown in Table 1).x_1 is uniformly distributed on [0,1] for Circle and Plane; [0,10] for SineV; and [0,4π] for SineH. x_2 is uniformly distributed on [0,1] for Circle and Plane; [-10,10] for SineV; and [0,10] for SineH. x_3 is uniformly distributed on [0,5] for Plane. Each dataset has 20,000 data instances, and concept drift occurs at 10,000. A drift is simulated by varying the parameter in each data set (as described in Table 1).

Experiment Setting and Results Analysis. The window sizes w_1 and w_2 are set to 1000 and 200, respectively, with sliding steps L of 20. In this series of tests, we do not employ PH test to detect drift, nor do we need to compare it to other algorithms, as our sole objective is to illustrate the behavior of reconstruction errors on changing data. Figure 3 shows the variation of the reconstruction error as S_{cur} slides over data streams from different datasets having different types of drifts.

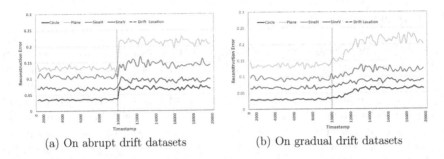

(a) On abrupt drift datasets   (b) On gradual drift datasets

Fig. 3. Reconstruction errors on abrupt drift and gradual drift datasets, respectively

It is readily apparent that the reconstruction error appears to remain practically constant until the drift point is reached. When abrupt concept drift happens (as depicted in Fig. 3(a)), the current window is filled with data relating to the new concept, and the reconstruction error increases abruptly. If the type of concept drift is gradual (as depicted in Fig. 3(b)), the number of sample points in the current window that correspond to the new concept increases gradually as fresh data come in. Consequently, the reconstruction error increases gradually with time. The experimental results demonstrate that reconstruction errors of SAE on the current windows can correctly characterize the distribution differences of the data.

Table 1. Datasets used for the experiments in Sect. 4.1

Dataset	Dim	Drift Type	Problems Description	Fixed Values	Before → After Drift
Circe [7,12,26]	2	abrupt	$(x_1 - a)^2 + (x_2 - b)^2 \leq r$	$a=b=0.5$	r:0.2→0.3
SineV [7,26]	2	abrupt	$x_2 \leq a\,(\sin(bx_1 + c)) + d$	$a=b=1,c=0$	d:-2→1
SineH [26]	2	abrupt	$x_2 \leq a\,(\sin(bx_1 + c)) + d$	$a=d=5,b=1$	c:0→-π
Plane [7,12]	3	abrupt	$x_3 \leq -a_0 + a_1x_1 + a_2x_2$	$a_1=a_2=0.1$	a_0:-1→-3.2
Circle(G)	2	gradual	Same as Circle except for drift type		
SineV(G)	2	gradual	Same as SineV except for drift type		
SineH(G)	2	gradual	Same as SineH except for drift type		
Plane(G)	3	gradual	Same as Plane except for drift type		

4.2 Evaluate the Detection Capability of DD-SAPH

Datasets. In this section we select four synthetic datasets with multiple change points to evaluate the performance of DD-SAPH on data streams containing multiple concept drifts. All four datasets are designed for the binary classification problem. By picking only samples from the positive class, which changes the classification issue into a change detection problem, we refer to Li Bu et al. [8]

- SINE2 [14]:It has 50k data samples. Classification function is $x_2 = 0.5 + 0.3\sin(3\pi x_1)$.
- Stagger [14]: It has 30k data samples; Under the first context, samples satisfying $size = small \land color = red$ are labeled as positive; Under the second context, samples satisfying $color = green \lor shape = circular$ are labeled as positive; Under the third context, samples satisfying $size = medium \lor size = large$ are labeled as positive.
- MIX [14]: It has 50k data samples. Samples satisfying $v = true$, $w = true$, $x_2 < 0.5 + 0.3\sin(3\pi x_1)$ are labeled as positive.
- SINE2G [3] has the same setting as SINE2 except the drift type.

A drift is simulated by changing or reversing the classification equation. To be specific, the classification is reversed when drift occurs in SINE2, SINE2G and MIX. The context is changed when drift occurs in Stagger. Drift occurs every 10000 samples in all datasets.

Evaluation Metrics and Comparing Algorithms. We introduce the same evaluation metrics as in [10], namely TP, FP, FN, and Late. TP is the number of drifts identified within an acceptable delay duration Δ, where Δ is defined as within $2w$ instances of the arrival of the new concept. w is the length of the current window. FP represents the number of drifts detected erroneously. FN represents the number of drift points that exist but have not been discovered. Late is the number of drifts detected outside of Δ. HDDDM [11], kdqTree [10], and D3 [15] are three unsupervised detection methods we will compare to our approach.

Experiment Setting and Results Analysis. The selection of parameters for kdqTree, HDDDM, and D3 adheres to the parameter-setting rules outlined in their individual papers. In kdqTree, $\delta = 0.01$, $k = 500$, and the window size is 200. In HDDDM, $N = 200$, $b = \lfloor\sqrt{N}\rfloor$, and $\alpha = 0.1$. In D3, $\rho = 0.1$, $\tau = 0.7$, and the window size is set to 200, in line with the other algorithms. In our method, set $w_1 = 1000$, $w_2 = 200$, which is consistent with the current window (data batch) size of the three comparing algorithms. Set the sliding steps L to 20 and the PH test parameters used to determine the dynamic threshold to $\delta = 0.0003$ and $\lambda = 5$.

The results of the experiment are presented in Table 2, demonstrating that kdqTree, HDDDM, and D3 fail to strike an optimal balance between sensitivity and robustness in the face of changes. More specifically, kdqTree is less sensitive to changes. In spite of the absence of false positives, the detection latency of kdqTree is lengthy and its missed detections are possible. KdqTree is unsuitable for cases with slight changes. HDDDM and D3 are overly sensitive to abrupt changes. Particularly, the number of false positives in HDDDM was ten times that of actual drift points. Despite having a short latency in detection, they

Table 2. Detection results of four unsupervised detection methods on data streams with multiple drift points

Methods	Dataset	TP	Late	FP	FN
DD-SAPH	SINE2	4	0	0	0
	MIX	4	0	0	0
	Stagger	1	1	0	0
	SINE2G	0	4	1	0
kdqTree	SINE2	0	4	0	0
	MIX	0	4	0	0
	Stagger	0	0	0	2
	SINE2G	0	4	0	0
HDDDM	SINE2	3	0	41	1
	MIX	4	0	56	0
	Stagger	2	0	19	0
	SINE2G	2	2	56	0
D3	SINE2	4	0	5	0
	MIX	4	0	8	0
	Stagger	2	0	4	0
	SINE2G	0	3	3	1

will produce an overwhelming amount of false positives. Therefore, they are not suitable for scenarios with a lot of noise. DD-SAPH's performance is the most superior, capturing every distribution change precisely and promptly while ensuring low false positives and low latency. In details, DD-SAPH exhibits a low false positive rate for abrupt drift, with over 92% of change points being detected within Δ. All change points can be detected within 2Δ after the change. For gradual drift, DD-SAPH ensures timely detection of any distribution change within 2Δ, while minimizing the number of false positives.

4.3 Comparison with Other Unsupervised Methods Under Classification Task

Datasets. One artificial dataset InterchangingRBF (D1) [23] and three real network datasets Sensor (D2) [6], Outdoor Stream (D3) [22] and Spambase (D4) [17] were selected for this part of the experiment:

Evaluation Metrics and Comparing Algorithms. The classification accuracy of the classifier is used as an evaluation metric to illustrate how DD-SAPH enhances the performance of the base classification model on these four datasets. When calculating classification accuracy, we assume that labels are immediately available. It is important to note that labels are only used to evaluate the algorithm's drift detection performance and have no effect on the drift algorithm.

Interleaved-Test-Then-Train (prequential) [5] is the evaluation strategy adopted; each newly arrived instance is classified before being used to train the classifier incrementally. The Hoeffding tree with default parameter settings within the scikit-multiflow platform is used as the base classifier in this experiment due to its efficacy and generalization. Initially, a number of data points are required to train the model; this phase is frequently referred to as the offline or warm-up phase of the classification model [1].

In addition to the three comparing algorithms described in Sect. 4.2 (kdqTree, HDDDM, and D3), a static baseline Without Detection (WD) is introduced. WD does not perform concept drift detection and adaptation. It is only evaluated using the prequential mentioned above. Specifically, the current classifier classifies each newly-arrived instance and stores the evaluation metric value. This instance is then used to incrementally update the current classifier.

Experiment Setting and Results Analysis. The parameter settings for the kdqTree, HDDDM, and D3 remain the same as in Sect. 4.2. To reduce the impact of the model's ignorance of the data stream on the classification accuracy in its initial state, 200 data points are used to warm up the classification model. For our detection method, we set the window size $w_1 = 1000$, $w_2 = 200$, and sliding steps $L = 20$. The PH test parameters for warning and alarming drifts were set as follows: $\delta = 0.0003$, $\lambda_d = 0.007$, and $\lambda_w = 0.003$. When the warning threshold is exceeded, DD-SAPH initiates and maintains an incremental training of a new classifier M_{new}. The old classifier is replaced with M_{new} and the reference window is reset to retrain SAE once the drift threshold is reached. Considering the inevitable delay in detection, the reference window retains $\frac{w_1}{2}$ old data. Due to the fact that no concept drift warning value is set in any of the other comparing methods, the classifier is reset whenever a concept drift is detected.

Table 3 presents the classification accuracy results of the Hoeffding tree after the addition of various unsupervised detection methods. Our experimental findings demonstrate the rapid adaptation of our proposed approach to changes in the data stream, resulting in improved classifier performance and generalizability across different dataset scenarios. Notably, the Hoeffding tree classifier combined with DD-SAPH outperforms the static baseline WD and other comparing methods in terms of classification accuracy across all dataset scenarios. It is worth mentioning that DD-SAPH will perform better on real datasets than on artificial datasets because, in the real world, $P(X)$ is the primary factor influencing $P(y|X)$.

Table 3. Classification accuracy of the Hoeffding tree after adding different unsupervised concept drift detection methods

Dataset	WD	kdqTree	HDDDM	D3	DD-SAPH
D1	25.80	25.80	89.25	58.71	**89.41**
D2	72.62	60.30	79.56	67.86	**84.76**
D3	59.65	60.31	55.61	59.61	**61.87**
D4	92.61	91.61	97.64	97.84	**98.86**

5   Conclusion and Future Work

In conclusion, this paper presents a new unsupervised concept drift detection algorithm, DD-SAPH, which utilizes the original features of data to detect concept drift in a data stream. Unlike existing methods, DD-SAPH does not rely on fixed statistical characteristics or nonparametric estimation methods. Instead, it utilizes a SAE to automatically extract hidden features that represent the data distributions, allowing for more adaptable and flexible estimation. This approach has the potential to improve the accuracy and reliability of concept drift detection in real-world scenarios with limited data samples. The reference window, which represents the initial concept, is divided up into numerous small windows by DD-SAPH. Each small window's samples are then transformed into training data to guide the training of SAE, which represents the initial distribution. As a metric, DD-SAPH uses the reconstruction error on the current window (which represents the current distribution) to measure the difference between the current distribution and the initial distribution. Adaptive thresholds are calculated dynamically by the PH test, and drift is flagged when the difference in distributions exceeds the threshold. DD-SAPH supports the use of any classifier that supports incremental updates and lacks a built-in module for concept drift detection.

Taking into account the influence of window size on detection performance, our next direction will be to investigate the adaptive adjustment of window size for various datasets.

Acknowledgments. This research was supported by the National Key Research and Development Program of China under Grant No. 2022ZD0115403, National Natural Science Foundation of China under Grant No.62072236, and the Fundamental Research Funds for the Central Universities under Grant NO.56XCA2205404.

References

1. Agrahari, S., Singh, A.K.: Concept drift detection in data stream mining: a literature review. J. King Saud Univ.-Comput. Inform. Sci. (2021)
2. Ahmadi, Z., Beigy, H.: Semi-supervised ensemble learning of data streams in the presence of concept drift. In: Corchado, E., Snášel, V., Abraham, A., Woźniak, M.,

Graña, M., Cho, S.-B. (eds.) HAIS 2012. LNCS (LNAI), vol. 7209, pp. 526–537. Springer, Heidelberg (2012). https://doi.org/10.1007/978-3-642-28931-6_50

3. Baena-Garcıa, M., del Campo-Ávila, J., Fidalgo, R., Bifet, A., Gavalda, R., Morales-Bueno, R.: Early drift detection method. In: Fourth International Workshop on Knowledge discovery from Data Streams, vol. 6, pp. 77–86 (2006)

4. Bifet, A., Gavalda, R.: Learning from time-changing data with adaptive windowing. In: Proceedings of the 2007 SIAM International Conference on Data Mining, pp. 443–448. SIAM (2007)

5. Bifet, A., et al.: Moa: Massive online analysis, a framework for stream classification and clustering. In: Proceedings of the First Workshop on Applications of Pattern Analysis, pp. 44–50. PMLR (2010)

6. Bodik, P., Hong, W., Guestrin, C., Madden, S., Paskin, M., Thibaux, R.: MIT sensor data. http://db.csail.mit.edu/labdata/labdata.html (2004)

7. Bu, L., Alippi, C., Zhao, D.: A pdf-free change detection test based on density difference estimation. IEEE Trans. Neural Netw. Learn. Syst. 29(2), 324–334 (2016)

8. Bu, L., Zhao, D., Alippi, C.: An incremental change detection test based on density difference estimation. IEEE Trans. Syst. Man Cybern. Syst. 47(10), 2714–2726 (2017)

9. Cerqueira, V., Gomes, H.M., Bifet, A., Torgo, L.: Studd: a student-teacher method for unsupervised concept drift detection. Mach. Learn. 1–28 (2022)

10. Dasu, T., Krishnan, S., Venkatasubramanian, S., Yi, K.: An information-theoretic approach to detecting changes in multi-dimensional data streams. In: Proceedings of Symposium on the Interface of Statistics, Computing Science, and Applications (Interface) (2006)

11. Ditzler, G., Polikar, R.: Hellinger distance based drift detection for nonstationary environments. In: 2011 IEEE Symposium on Computational Intelligence in Dynamic and Uncertain Environments (CIDUE), pp. 41–48. IEEE (2011)

12. Elwell, R., Polikar, R.: Incremental learning of concept drift in nonstationary environments. IEEE Trans. Neural Netw. 22(10), 1517–1531 (2011)

13. Frias-Blanco, I., del Campo-Ávila, J., Ramos-Jimenez, G., Morales-Bueno, R., Ortiz-Diaz, A., Caballero-Mota, Y.: Online and non-parametric drift detection methods based on hoeffding's bounds. IEEE Trans. Knowl. Data Eng. 27(3), 810–823 (2014)

14. Gama, J., Medas, P., Castillo, G., Rodrigues, P.: Learning with drift detection. In: Bazzan, A.L.C., Labidi, S. (eds.) SBIA 2004. LNCS (LNAI), vol. 3171, pp. 286–295. Springer, Heidelberg (2004). https://doi.org/10.1007/978-3-540-28645-5_29

15. Gözüaçık, Ö., Büyükçakır, A., Bonab, H., Can, F.: Unsupervised concept drift detection with a discriminative classifier. In: Proceedings of the 28th Acm International Conference on Information and Knowledge Management, pp. 2365–2368 (2019)

16. Haque, A., Khan, L., Baron, M.: Sand: semi-supervised adaptive novel class detection and classification over data stream. In: Proceedings of the AAAI Conference on Artificial Intelligence, vol. 30 (2016)

17. Hopkins, M., Reeber, E., Forman, G., Suermondt, J.: UCI Machine Learning Repository - Spambase Dataset (1999). http://archive.ics.uci.edu/ml/datasets/Spambase

18. Hosseini, M.J., Gholipour, A., Beigy, H.: An ensemble of cluster-based classifiers for semi-supervised classification of non-stationary data streams. Knowl. Inf. Syst. 46(3), 567–597 (2016)

19. Kifer, D., Ben-David, S., Gehrke, J.: Detecting change in data streams. In: VLDB, Toronto, Canada, vol. 4, pp. 180–191 (2004)

20. Liu, A., Lu, J., Liu, F., Zhang, G.: Accumulating regional density dissimilarity for concept drift detection in data streams. Pattern Recogn. **76**, 256–272 (2018)
21. Liu, A., Song, Y., Zhang, G., Lu, J.: Regional concept drift detection and density synchronized drift adaptation. In: IJCAI International Joint Conference on Artificial Intelligence (2017)
22. Losing, V., Hammer, B., Wersing, H.: Interactive online learning for obstacle classification on a mobile robot. In: 2015 International Joint Conference on Neural Networks (ijcnn), pp. 1–8. IEEE (2015)
23. Losing, V., Hammer, B., Wersing, H.: Knn classifier with self adjusting memory for heterogeneous concept drift. In: 2016 IEEE 16th International Conference on Data Mining (ICDM), pp. 291–300. IEEE (2016)
24. Lu, J., Liu, A., Dong, F., Gu, F., Gama, J., Zhang, G.: Learning under concept drift: A review. IEEE Trans. Knowl. Data Eng. **31**(12), 2346–2363 (2018)
25. Lu, N., Zhang, G., Lu, J.: Concept drift detection via competence models. Artif. Intell. **209**, 11–28 (2014)
26. Minku, L.L., White, A.P., Yao, X.: The impact of diversity on online ensemble learning in the presence of concept drift. IEEE Trans. Knowl. Data Eng. **22**(5), 730–742 (2009)
27. Qahtan, A.A., Alharbi, B., Wang, S., Zhang, X.: A pca-based change detection framework for multidimensional data streams: change detection in multidimensional data streams. In: Proceedings of the 21th ACM SIGKDD International Conference on Knowledge Discovery and Data Mining, pp. 935–944 (2015)
28. Raab, C., Heusinger, M., Schleif, F.M.: Reactive soft prototype computing for concept drift streams. Neurocomputing **416**, 340–351 (2020)
29. Ross, G.J., Tasoulis, D.K., Adams, N.M.: Nonparametric monitoring of data streams for changes in location and scale. Technometrics **53**(4), 379–389 (2011)
30. Sethi, T.S., Kantardzic, M.: On the reliable detection of concept drift from streaming unlabeled data. Expert Syst. Appl. **82**, 77–99 (2017)
31. Sobhani, P., Beigy, H.: New drift detection method for data streams. In: Bouchachia, A. (ed.) ICAIS 2011. LNCS (LNAI), vol. 6943, pp. 88–97. Springer, Heidelberg (2011). https://doi.org/10.1007/978-3-642-23857-4_12
32. Tibshirani, R.J., Efron, B.: An introduction to the bootstrap. Monographs Stat. Appli. Probabil. **57**, 1–436 (1993)

An Industrial Robot Path Planning Method Based on Improved Whale Optimization Algorithm

Peixin Huang[1,2], Chen Dong[1,2]([✉]), Zhenyi Chen[3], Zihang Zhen[4], and Lei Jiang[4]

[1] College of Computer and Data Science, Fuzhou University, Fuzhou 350116, China
dongchen@fzu.edu.cn
[2] Fujian Key Laboratory of Network Computing and Intelligent Information Processing (Fuzhou University), Fuzhou, China
[3] Department of Computer Science and Engineering, University of South Florida, Tampa, FL 33620, USA
zhenyichen@usf.edu
[4] College of Computer and Cyber Security, Fujian Normal University, Fuzhou 350007, China

Abstract. With the development of technology, robots are gradually being used more and more widely in various fields. Industrial robots need to perform path planning in the course of their tasks, but there is still a lack of a simple and effective method to implement path planning in complex industrial scenarios. In this paper, an improved whale optimization algorithm is proposed to solve the robot path planning problem. The algorithm initially uses a logistic chaotic mapping approach for population initialization to enhance the initial population diversity, and proposes a jumping mechanism to help the population jump out of the local optimum and enhance the global search capability of the population. The proposed algorithm is tested on 12 complex test functions and the experimental results show that the improved algorithm achieves the best results in several test functions. The algorithm is then applied to a path planning problem and the results show that the algorithm can help the robot to perform correct and efficient path planning.

Keywords: Robot path planning · Whale optimization algorithm · Chaotic mapping · Jumping mechanism

1 Introduction

With the continuous development of Artificial Intelligence (AI), science and technology, smart factories are gradually going digital [1]. AI is already widely used [2] and robots are becoming a sign of the times and have been widely used in industry [3], agriculture [4], military [5], medical [6] and other fields. With the development of robotics, mobile robots have gradually derived different forms, such as

H. Jin et al. (Eds.): GPC 2023, LNCS 14503, pp. 209–222, 2024.
https://doi.org/10.1007/978-981-99-9893-7_16

unmanned aerial vehicles (UAVs), automatic guided vehicles (AGVs), unmanned surface vessels (USVs), automatic underwater vehicles (AUVs) [7], etc. Intelligent robots are gradually becoming an essential part of Industry 4.0 [8].

Industrial robots move during their tasks, during which they may encounter obstacles and therefore need to perform path planning [9]. The robot path planning problem is to find a path from the starting point to the target point with the goal of being collision-free and having the shortest possible path length [10]. The robot path planning problem is an NP-hard problem, especially when the environment becomes more complex, and the difficulty of its solution will increase significantly [11].

To meet the robot's needs for path planning, an improved whale optimization algorithm is proposed to solve the robot path planning problem. A logistic chaotic mapping method is used for population initialization of the whale optimization algorithm, and a jumping mechanism is introduced to enhance the global search capability of the algorithm. And then the improved algorithm is used for path planning to solve the robot path planning problem. The main contributions of this paper can be summarized as follows:

1. Considering the problem that the population initialization is more random and unevenly distributed, a logistic chaotic mapping method is used for population initialization to improve the diversity of the initial population so that the population can be as close as possible to the potential optimal solution at the early stage of the algorithm.
2. To address the problem that the algorithm is prone to fall into local optimum, a jumping mechanism is proposed to retain a portion of the elite group with good fitness values, while the remaining group is jumped to find potential optimal solutions, thus avoiding the problem of falling into local optimum while strengthening the global search capability of the algorithm.
3. In order to use the proposed improved whale optimization algorithm for solving robot path planning problems, the path planning problem is modeled to apply the method, and the effectiveness of the algorithm in solving complex path planning problems for industrial scenarios is confirmed experimentally.

2   Related Work

In recent years, many experts and scholars have explored in robot path planning and swarm intelligence optimization methods. This section demonstrates the relevant explorations by experts and scholars.

2.1   Robot Path Planning

The robot path planning problem is very complex, and experts and scholars have tried different approaches to try to solve the problem. Chen et al. [12] proposed an improved RRT-Connect mobile robot path planning algorithm that improves the detection efficiency of the algorithm by generating a simple and efficient third node in space, allowing the algorithm to expand greedily using a quadtree, and adding bootstrapping to the algorithm so that it features an offset towards the target point when expanding new nodes. Li et al. [13] proposed an improved PRM algorithm that optimizes the generation of sampling points, removes redundant sampling points, sets a distance threshold between road points, and uses a two-way incremental collision detection method to reduce the number of collision detection calls and improve the roadmap construction efficiency. Semnani et al. [14] proposed a hybrid algorithm combining deep reinforcement learning with the FMP algorithm, which improves path planning success by using the FMP algorithm in simple situations, crowded situations and when the robot is stuck, and by continuing to use deep reinforcement learning in normal situations where the FMP algorithm does not produce an optimal path. Lai et al. [15] proposed a centrally constrained weighted A* algorithm, which designed dynamic weights to assign different dynamic weights to nodes at different locations so as to control the direction of node exploration expansion, and added adaptive thresholds to the heuristic function to adaptively change the weights of the heuristic function in real time according to the map size to enhance the adaptive capability of the algorithm and improve the search efficiency. Huang et al. [16] proposed an improved APF method based on parallel search, introducing a method to bypass the nearest obstacle, thus helping the UAV to successfully escape the local minimum without colliding with any obstacle, and proposed the idea of parallel search to solve the target unreachability problem of the traditional APF method when there are too many obstacles.

2.2   Swarm Intelligence Optimization

As the problem of robot path planning has been studied by experts and scholars, swarm intelligence algorithms have come to the forefront of their attention for their ease of implementation and excellent performance. Gao et al. [17] proposed an improved heuristic ACO algorithm which considered the distance from the ant's current position to the target point and the heuristic distance from the ant's nearest neighbour to the target point, and devised a pheromone diffusion gradient formula, as well as introduced a backtracking strategy and a path merging strategy to further obtain the optimal path. Miao et al. [18] proposed an improved adaptive ACO that introduces an angle guidance factor and an obstacle exclusion factor for the robot path planning problem, and introduces a heuristic information adaptive adjustment factor and an adaptive pheromone volatility factor in the pheromone update rule to balance the convergence and global search capability of the ACO. Song et al. [19] proposed an improved PSO algorithm that applies certain perturbations to the swarm according to its evolutionary state, thus improving the ability of the swarm to jump out of local minima and combining it with Bessel curves to achieve smooth path control of the robot. Zhang et al. [20] proposed a hybrid algorithm based on the genetic algorithm and the firefly algorithm, in which the locally optimal fireflies are considered as a population when the firefly algorithm falls into a local optimal solution, and the population is subjected to the selection, crossover and mutation operations of the genetic algorithm, and the optimal firefly individuals are further obtained through genetic operations. Garcia et al. [21] proposed a multi-robot path planning method that combines the optimization capabilities of the A* algorithm with the search capabilities of co-evolutionary algorithms to plan collision-free paths in advance, making it unnecessary to consider obstacle avoidance once the robot has started running.

2.3   Motivation

In the recent research literature related to swarm intelligence and path planning, whale optimization algorithms are less frequently addressed. However, as a new swarm intelligence algorithm with excellent solving ability, it is still of great academic value to study the whale optimization algorithm. Among the methods for solving robot path planning problems introduced in the literature, there is still a lack of a simple and proven method for implementing it in complex industrial scenarios. Therefore, this paper investigates the whale optimization algorithm and its improvement and applies it to the path planning problem for path planning problem solving.

3   Improved Whale Optimization Algorithm

Whale Optimization Algorithm (WOA) is a novel nature-inspired meta-heuristic optimization algorithm, which is proposed in 2016 [22]. The literature states that WOA shows sufficient competitiveness against other state-of-the-art meta-heuristics methods. In this paper, logistic chaotic mapping method and jumping mechanism are introduced to enhance the performance of WOA.

3.1   The Classic Algorithm

The algorithm is inspired by the social behavior of humpback whales and simulates the special hunting method of humpback whales. The foraging behavior of humpback whales consists of three main mechanisms: (1) encircling prey, (2) bubble-net attacking and (3) search for prey.

Encircling Prey. The algorithm considers the current population-optimal solution as the prey, and the individuals other than the optimal one try to update their positions toward the prey, and this behavior is represented by Eqs. 1 and 2:

$$\overrightarrow{D_1} = \left| \overrightarrow{C} \cdot \overrightarrow{X^*}(t) - \overrightarrow{X}(t) \right| \tag{1}$$

$$\overrightarrow{X}(t+1) = \overrightarrow{X^*}(t) - \overrightarrow{A} \cdot \overrightarrow{D_1} \tag{2}$$

where t denotes the current iteration, $\overrightarrow{A}$ and $\overrightarrow{C}$ are coefficient vectors, $\overrightarrow{X^*}(t)$ denotes the current optimal solution, and $\overrightarrow{X}(t)$ denotes the current individual.

The coefficient vectors $\vec{A}$ and $\vec{C}$ are calculated by Eqs. 3 and 4:

$$\vec{A} = 2\vec{a} \cdot \vec{r_1} - \vec{a} \tag{3}$$

$$\vec{C} = 2 \cdot \vec{r_2} \tag{4}$$

where $\vec{r_1}$, $\vec{r_2}$ are random vectors of each dimension belonging to [0,1]. $\vec{a}$ decreases linearly during the iterative process and the $\vec{a}$ is calculated as shown in Eq. 5:

$$a = a_{max} - \frac{t \cdot a_{max}}{t_{max}} \tag{5}$$

where a_{max} denotes the maximum value of a, which is the initial value, and t_{max} denotes the maximum number of iterations.

Bubble-Net Attacking. When a humpback whale finds its prey, it will approach and attack in a spiral fashion with the position update equation shown in Eqs. 6 and 7:

$$\vec{D_2} = \left| \vec{X^*}(t) - \vec{X}(t) \right| \tag{6}$$

$$\vec{X}(t+1) = \vec{D_2} \cdot e^{bl} \cdot \cos(2\pi l) + \vec{X^*}(t) \tag{7}$$

where b is a constant that defines the shape of the logarithmic spiral and l is a random number in $[-1,1]$.

Whales will approach their prey in a spiral and some will swim away to surround their prey, assuming a probability of 50% for each of these two behaviors, the equation for whale position update is shown in Eq. 8:

$$\vec{X}(t+1) = \begin{cases} \vec{X^*}(t) - \vec{A} \cdot \vec{D_1} & p < 0.5 \\ \vec{D_2} \cdot e^{bl} \cdot \cos(2\pi l) + \vec{X^*}(t) & p \geq 0.5 \end{cases} \tag{8}$$

where p is a random number within [0,1].

Search for Prey. When $0 \leq \left| \vec{A} \right| \leq 1$, humpback whales go to encircle and feed on prey, while when $1 < \left| \vec{A} \right| \leq 2$, humpback whales randomly search for prey based on the population's position in relation to each other, and the position update formula for this behavior is calculated by Eqs. 9 and 10:

$$\vec{D_3} = \left| \vec{C} \cdot \vec{X_{rand}} - \vec{X}(t) \right| \tag{9}$$

$$\vec{X}(t+1) = \vec{X_{rand}} - \vec{A} \cdot \vec{D_3} \tag{10}$$

where $\vec{X_{rand}}$ denotes the position vector of a random whale in the current population.

3.2   Chaotic Mapping Initialization

The classic WOA uses a random distribution of whale populations at the beginning of the algorithm, which may lead to an uneven initial population distribution and poor population diversity, which in turn reduces the search performance of the algorithm. In this paper, a logistic chaotic mapping method is used to initialize the population to enhance the diversity of the initial population.

Scientific studies have shown that chaotic mappings have the property of being highly dependent on initial values, and two similar initial values also yield completely different sequences of random numbers. The use of chaotic mappings can yield better initial populations than uniform random distributions of diversity [23]. This property brings great benefits to the optimization calculation. The variable update formula of the logistic chaotic mapping used in this paper is calculated by Eq. 11:

$$x_{n+1} = \mu x_n (1 - x_n) \tag{11}$$

where μ is a control parameter within [0,4] and x_n takes values in the range $0 \leq x_n \leq 1$.

3.3   Jumping Mechanism

Although the classic WOA has a certain degree of random search mechanism, it may still fall into local optimum. In order to solve the problem of classic WOA falling into local optimum, a jumping mechanism is proposed to keep a part of the elite group with good fitness values when the algorithm falls into local optimum, while the rest of the group makes jumps to find possible more optimal solutions, thus avoiding the problem of falling into local optimum and enhancing the global search capability of the algorithm at the same time. The jumping mechanism is described in detail as follows:

The condition for implementing the jumping mechanism is that the optimal solution of the algorithm in k consecutive iterations all satisfy the following Eq. 12:

$$\left| \frac{f_{current\_best}}{f_{global\_best}} \right| < \varepsilon \tag{12}$$

That is, if the optimal solution of the algorithm changes by less than some threshold ε in several consecutive iterations, it is considered that the algorithm may be trapped in a local optimum and the jumping mechanism starts to be implemented at this point.

(1) A certain percentage of the elite group is retained, and this percentage is set to r.
(2) For each individual waiting to jump, a random vector $\vec{j}$ with each dimension between [0,1] is generated.
(3) Based on the random vector $\vec{j}$, the individual makes a jump whose position update formula is calculated by Eq. 13:

Table 1. Description of test functions.

Function	Range	f_{min}
$f_1(x,y) = 0.5 + \frac{\sin^2\left(x^2-y^2\right)-0.5}{\left[1+0.001\left(x^2+y^2\right)\right]^2}$	[−100,100]	0
$f_2(x,y) = -(y+47)\sin\sqrt{\left\|\frac{x}{2}+(y+47)\right\|} - x\sin\sqrt{\|x-(y+47)\|}$	[−512,512]	−959.6407
$f_3(x,y) = (x+2y-7)^2 + (2x+y-5)^2$	[−10,10]	0
$f_4(x,y) = 0.26\left(x^2+y^2\right) - 0.48xy$	[−10,10]	0
$f_5(x,y) = -0.0001\left[\left\|\sin x \sin y \exp\left(\left\|100-\frac{\sqrt{x^2+y^2}}{\pi}\right\|\right)\right\|+1\right]^{0.1}$	[−10,10]	−2.06261
$f_6(x,y) = \sin^2 3\pi x + (x-1)^2\left(1+\sin^2 3\pi y\right) + (y-1)^2\left(1+\sin^2 2\pi y\right)$	[−10,10]	0
$f_7(x,y) = -20\exp\left[-0.2\sqrt{0.5\left(x^2+y^2\right)}\right] - \exp\left[0.5\left(\cos 2\pi x + \cos 2\pi y\right)\right] + e + 20$	[−5,5]	0
$f_8(x,y) = (1.5-x+xy)^2 + \left(2.25-x+xy^2\right)^2 + \left(2.625-x+xy^3\right)^2$	[−5,5]	0
$f_9(x,y) = -\left\|\sin x \cos y \exp\left(\left\|1-\frac{\sqrt{x^2+y^2}}{\pi}\right\|\right)\right\|$	[−10,10]	−19.2085
$f_{10}(x,y) = \left(x^2+y-11\right)^2 + \left(x+y^2-7\right)^2$	[−5,5]	0
$f_{11}(x,y) = 10 + x^2 + y^2 - 10\left(\cos 2\pi x + \cos 2\pi y\right)$	[−5.12,5.12]	0
$f_{12}(x,y) = 4x^2 - 2.1x^4 + \frac{x^6}{3} + xy - 4y^2 + 4y^4$	[−5,5]	−1.0316

$$\vec{X}(t) = (0.5\cdot\vec{e}+\vec{j})\cdot\vec{X}(t) \qquad (13)$$

where $\vec{e}$ denotes the unit vector.

Let the number of individuals in the population be N, and the individuals in the population are ranked according to their fitness values from good to bad, and let $Sequence_i$ be the ranking of individual i in that population, then the formula for updating the position of an individual after performing the jumping mechanism is calculated by Eq. 14:

$$\vec{X}(t) = \begin{cases} \vec{X}(t) & Sequence_i \leq r\cdot N \\ (0.5\cdot\vec{e}+\vec{j})\cdot\vec{X}(t) & others \end{cases} \qquad (14)$$

4 Experiments and Analysis About the Proposed Algorithm

In this section, 12 test functions are used to perform ablation experiments on the proposed algorithm to check the performance of the algorithm.

4.1 Test Functions

To test the solution performance of the algorithm, 12 complex functions were chosen, and the information of the functions is shown in Table 1. The images of these functions is shown in Fig. 1.

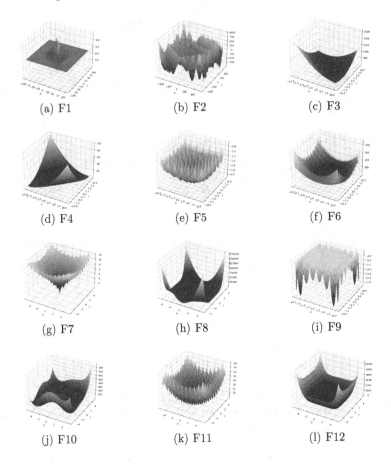

(a) F1   (b) F2   (c) F3

(d) F4   (e) F5   (f) F6

(g) F7   (h) F8   (i) F9

(j) F10   (k) F11   (l) F12

Fig. 1. Images of test functions.

4.2 Ablation Experiments

To check the performance of the algorithm, ablation experiments were performed. For each test function, it was run 1000 times, while the mean and standard deviation of the obtained results were taken afterwards for comparison. The experimental results are shown in Table 2.

As can be seen from Table 2, the proposed algorithm combining chaotic mapping and jumping mechanism achieves the best mean and standard deviation on most of the tested functions. In addition, we choose the convergence curves of F2 and F7 in one random run for comparison. The convergence curves of the four algorithms on F2 and F7 are shown in Figs. 2 and 3, respectively.

From Fig. 2, it is illustrated that the classic algorithm falls into a local optimum for a considerable period of time and fails to jump out even at the end. The algorithm with the introduction of the jumping mechanism, on the other hand, has also fallen into the local optimum, but it can still jump out of the local

Table 2. Experimental results of the mean and standard deviation of the algorithm on the test functions.

F	classic		chaotic		jump		chaotic+jump	
	mean	std	mean	std	mean	std	mean	std
F1	0.026848	0.032141	0.025892	0.030790	0.024879	0.030124	0.024794	0.028150
F2	−945.760	34.06391	−956.535	15.77971	−947.494	30.04956	−957.428	12.20646
F3	0.000580	0.000679	0.000613	0.000751	0.000644	0.000754	0.000612	0.000740
F4	3.39E-08	1.51E-07	3.28E-08	1.32E-07	4.07E-08	2.78E-07	3.95E-08	1.71E-07
F5	−2.06248	2.00E-04	−2.06249	1.64E-04	−2.06247	1.83E-04	−2.06248	1.70E-04
F6	0.001184	0.005164	0.001228	0.005250	0.001124	0.003759	0.001270	0.006145
F7	0.134977	0.545384	0.109590	0.488608	0.131709	0.539178	0.091034	0.436451
F8	0.034449	0.156929	0.028315	0.142711	0.038123	0.165445	0.026680	0.139056
F9	−19.0965	0.203425	−19.1049	0.149970	−19.0989	0.128065	−19.1055	0.124596
F10	0.009742	0.014397	0.008483	0.014228	0.009024	0.014020	0.008082	0.011427
F11	0.543029	0.587399	0.554303	0.594238	0.525165	0.564015	0.542261	0.593994
F12	−1.02005	0.092915	−1.02412	0.073140	−1.02251	0.082124	−1.02498	0.068634

optimum with the help of the jumping mechanism and finally find the optimal solution. The algorithm that introduces chaotic initialization, on the other hand, finds the optimal solution quickly because the initial diversity is better and it is closer to the optimal solution at the beginning.

From Fig. 3, it can be seen that the classic algorithm eventually falls into a local optimum that is difficult to jump out of, while the jumping mechanism can jump out of the local optimum to find the optimal solution. For the algorithm that introduces chaotic initialization, thanks to the better initial diversity, it maintains a more robust performance in the operation of the algorithm and is not limited by the local optimum, and still finds the optimal solution in the end.

5   Robot Path Planning Based on IWOA

In this section, the path planning problem is modeled and the proposed IWOA algorithm is applied to solve it.

5.1   Problem Modeling

Figure 4 illustrates a path from the starting point to the target point. The bottom left corner is the starting point of the path planning, and the top right corner is the target point of the path planning. The coordinate axes are divided equally into $n + 1$ copies. Let the coordinates of the start point be $(x_0, y_0) = (0,0)$ and the coordinates of the target point be (x_{n+1}, y_{n+1}). Thus, a point can be taken on each of the n partition lines and noted as $(x_1, y_1), (x_2, y_2), (x_3, y_3), ..., (x_n, y_n)$ in turn, and then a path can be obtained by connecting these points in turn. In other words, a set of points can uniquely correspond to the path of a robot. Thus,

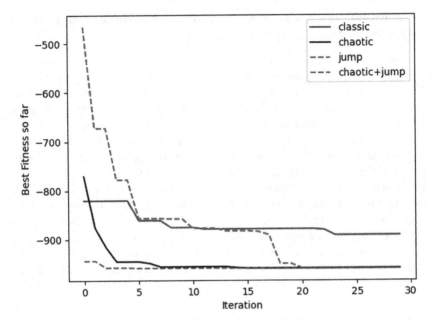

Fig. 2. The convergence curves on F2.

the path planning problem is transformed into an n-dimensional optimization problem. In practical application scenarios, the precision and smoothness of the path can be adjusted indirectly by adjusting the number of points.

For robot path planning problems, consideration should be given to making the path length as short as possible while avoiding collisions. Let the cost of a path be J_{path}. The cost J_{path} is calculated by Eq. 15:

$$J_{path} = J_{fuel} + J_{collision} \tag{15}$$

where J_{fuel} represents the path length, as calculated by Eq. 16:

$$J_{fuel} = \sum_{i=0}^{n} \sqrt{(x_{i+1} - x_i)^2 + (y_{i+1} - y_i)^2} \tag{16}$$

Considering collisions, the line segment joining (x_i, y_i) and (x_{i+1}, y_{i+1}) is denoted ℓ_i. Let there be m obstacles in the scene and the cost factor for each collision of the robot is λ. Then $J_{collision}$ is calculated by Eq. 17:

$$J_{collision} = \sum_{i=0}^{n} \sum_{j=1}^{m} \lambda \cdot G(\ell_i, obstacle_j) \tag{17}$$

where $G(\ell_i, obstacle_j)$ is calculated by Eq. 18:

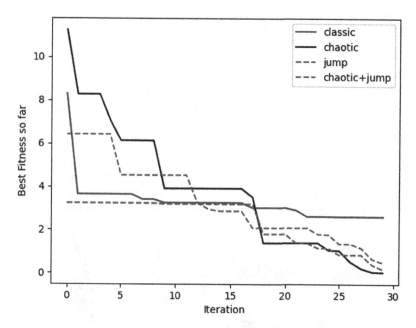

Fig. 3. The convergence curves on F7.

$$G(\ell_i, obstacle_j) = \begin{cases} 1 & \ell_i \cap obstacle_j \neq \emptyset \\ 0 & \ell_i \cap obstacle_j = \emptyset \end{cases} \qquad (18)$$

5.2 Experiments and Results Analysis

The proposed IWOA algorithm was used to solve a path planning problem and tested in a path planning scenario, the path planning results of which are shown in Fig. 5. It can be seen that the robot is able to plan the path from the starting point to the target point correctly.

Figure 6 illustrates the cost curve of the path planning process. It can be seen that early on in the algorithm, with the help of chaotic initialization, the initial diversity is better and the curve falls quickly, although there is a period in the middle of the algorithm when it falls into a local optimum, but with the help of the jumping mechanism, it still helps the algorithm to eventually jump out of the local optimum and find the optimal solution. Therefore, the proposed algorithm is able to solve the robot path planning problem effectively.

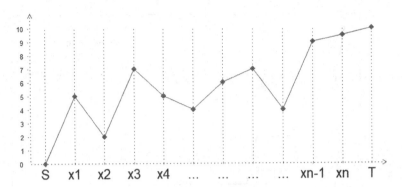

Fig. 4. Schematic diagram of robot path planning modeling.

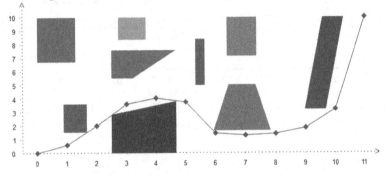

Fig. 5. Experimental results of path planning for an industrial scenario.

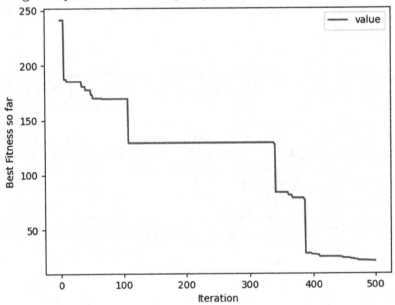

Fig. 6. Path planning process cost change curve.

6   Conclusion

An improved whale optimization algorithm is proposed to solve the robot path problem. A logistic chaos mapping method is introduced for population initialization to enhance the initial population diversity in order to address the problem of uneven initial population distribution in classic WOA. To address the problem that the classic WOA may fall into a local optimum, a jumping mechanism is proposed to retain a portion of the elite population and make the remaining individuals jump in order to try to jump out of the local optimum and try to find a better solution. Finally, the path planning problem is re-modelled and solved by applying the proposed algorithm. The results show that the algorithm proposed in this paper can effectively solve the robot path planning problem. In future research, path smoothness and dynamic obstacles will become further considerations and further solve more complex multi-robot path planning problems.

Acknowledgements. This work is supported by the fund of Fujian Province Digital Economy Alliance, the National Natural Science Foundation of China (No. U1905211), and the Natural Science Foundation of Fujian Province (No. 2020J01500).

References

1. Hong, Q., et al.: A dynamic demand-driven smart manufacturing for mass individualization production. In: 2021 IEEE International Conference on Systems, Man, and Cybernetics (SMC), pp. 3297–3302. IEEE (2021)
2. Dong, C., et al.: A cost-driven method for deep-learning-based hardware trojan detection. Sensors **23**(12), 5503 (2023)
3. Chen, Z., et al.: 6G mobile communications for multi-robot smart factory. J. ICT Stand. **9**, 371–404 (2021)
4. Fountas, S., et al.: Agricultural robotics for field operations. Sensors **20**(9), 2672 (2020)
5. Bistron, M., Piotrowski, Z.: Artificial intelligence applications in military systems and their influence on sense of security of citizens. Electronics **10**(7), 871 (2021)
6. Dupont, P.E., et al.: A decade retrospective of medical robotics research from 2010 to 2020. Sci. Robot. **6**(60), eabi8017 (2021)
7. Nair, V.G., Guruprasad, K.R.: 2 D-VPC: an efficient coverage algorithm for multiple autonomous vehicles. Int. J. Control Autom. Syst. **19**(8), 2891–2901 (2021)
8. Dong, C., et al.: A dynamic distributed edge-cloud manufacturing with improved ADMM algorithms for mass personalization production. J. King Saud Univ. Comput. Inf. Sci. **35**(8), 101632 (2023)
9. Yang, Y., Juntao, L., Lingling, P.: Multi-robot path planning based on a deep reinforcement learning DQN algorithm. CAAI Trans. Intell. Technol. **5**(3), 177–183 (2020)
10. Zheng, Q., et al.: An improved deep reinforcement learning for robot navigation. In: Third International Conference on Machine Learning and Computer Application (ICMLCA 2022), vol. 12636, pp. 171–176. SPIE (2023)
11. Kyprianou, G., Doitsidis, L., Chatzichristofis, S.A.: Towards the achievement of path planning with multi-robot systems in dynamic environments. J. Intell. Robot. Syst. **104**(1), 15 (2022)

12. Chen, J., Zhao, Y., Xing, X.: Improved RRT-connect based path planning algorithm for mobile robots. IEEE Access **9**, 145988–145999 (2021)
13. Li, Q., et al.: Smart vehicle path planning based on modified PRM algorithm. Sensors **22**(17), 6581 (2022)
14. Semnani, S.H., et al.: Multi-agent motion planning for dense and dynamic environments via deep reinforcement learning. IEEE Robot. Autom. Lett. **5**(2), 3221–3226 (2020)
15. Lai, X., Li, J.H., Chambers, J.: Enhanced center constraint weighted a* algorithm for path planning of petrochemical inspection robot. J. Intell. Robot. Syst. **102**, 1–15 (2021)
16. Huang, T., et al.: Path planning and control of a quadrotor UAV based on an improved APF using parallel search. Int. J. Aerosp. Eng. **2021**, 1–14 (2021)
17. Gao, W., et al.: An enhanced heuristic ant colony optimization for mobile robot path planning. Soft. Comput. **24**, 6139–6150 (2020)
18. Miao, C., et al.: Path planning optimization of indoor mobile robot based on adaptive ant colony algorithm. Comput. Ind. Eng. **156**, 107230 (2021)
19. Song, B., Wang, Z., Zou, L.: An improved PSO algorithm for smooth path planning of mobile robots using continuous high-degree Bezier curve. Appl. Soft Comput. **100**, 106960 (2021)
20. Zhang, T.-W., et al.: A new hybrid algorithm for path planning of mobile robot. J. Supercomputing **78**(3), 4158–4181 (2022)
21. García, E., et al.: An efficient multi-robot path planning solution using A* and coevolutionary algorithms. Integr. Comput. Aided Eng. **30**(1), 41–52 (2023)
22. Mirjalili, S., Lewis, A.: The whale optimization algorithm. Adv. Eng. Softw. **95**, 51–67 (2016)
23. Wang, X., Guan, N., Yang, J.: Image encryption algorithm with random scrambling based on one-dimensional logistic selfembedding chaotic map. Chaos, Solitons Fractals **150**, 111117 (2021)

Intrusion Detection System Based on Adversarial Domain Adaptation Algorithm

Jiahui Fei[1], Yunpeng Sun[1], Yuejin Wang[2], and Zhichao Lian[1(✉)]

[1] School of Cyberspace Security, Nanjing University of Science and Technology, Nanjing, China
lzcts@163.com
[2] School of Computing, Nanjing University of Science and Technology, Nanjing, China

Abstract. With the explosive growth of the Internet, massive high-dimensional data and multiple attack types make intrusion detection systems face greater challenges. In practical application scenarios, the amount of abnormal data is small, and intrusion detection systems in different scenarios cannot be quickly migrated, and specific intrusion detection systems need to be trained for different scenarios, which greatly wastes manpower and material resources. Therefore, in view of the hierarchical characteristics of network data streams, this paper uses CNN and RNN networks to extract the spatiotemporal features of network data streams, then input them into GAN for unsupervised learning. Considering that long and short-term recurrent neural network (LSTM-RNN) has been shown to be able to obtain information and learn complex time series by remembering the backward (or even forward) time steps of cells, this paper replaces the generator and discriminator of GAN with LSTM-RNN. Anomaly detection is then performed based on residual loss and identification loss. Finally, this paper uses the deep domain adaptation algorithm to map the target domain and the source domain, and then optimizes the confusion loss of the domain by adversarial training, and finally extracts the invariant features of the target and the source domain.

Keywords: Intrusion Detection System (IDS) · LSTM-RNN · Adversarial Domain Algorithm

1 Introduction

The Internet plays an irreplaceable role in finance, politics, culture, communications and other fields, as an important part of the new era and new infrastructure, it is also one of China's key information infrastructure. However, as people's lives and networks gradually converge, the frequency of cyberattacks is increasing year by year. According to the report, in 2021 alone, there are more than 1.19 million malicious program terminals infected with Trojans or botnets in China, and the number of websites that have been tampered with, implanted with backdoors and counterfeited is more than 22,000, according to the data of the National Information Security Vulnerability Sharing Platform (CNVD), China's network security vulnerabilities are as high as 1,600, of which 570 are high-risk vulnerabilities, and China's form in the Internet field is still very serious. China must also come up with practical ways to deal with the endless cyber attacks.

Intrusion detection system [1] (IDS), the concept was first proposed in 1980, which is a network security prevention and control technology that distinguishes benign traffic and malicious traffic during network transmission through real-time monitoring system. In recent years, with the development of machine learning (ML), the application of artificial intelligence technology in the field of intrusion detection has gradually increased.

However, while much previous work has had some success in developing intrusion detection systems, with the increasing level and complexity of modern network connectivity, the resource cost of acquiring network traffic and labels in different environments and designing intrusion detection systems in a targeted manner is significant. This places high demands on the portability of intrusion detection systems.

Therefore, an intrusion detection system based on adversarial domain adaptation algorithm is proposed. On the basis of transfer learning, aiming at the problem that intrusion detection models in different scenarios are not easy to migrate and transfer learning efficiency is poor, the feature extraction cascade of CNN and RNN is first used to extract the multi-dimensional spatiotemporal features of the network flow, and then the generator and discriminator of GAN are replaced with LSTM-RNN for adversarial traffic generation using anomaly detection based on adversarial training, and finally the transfer learning algorithm based on adversarial domain adaptation is used to project the target domain and the source domain into the same feature space for feature transfer.

The contributions of this article are as follows:

- This paper proposes a network traffic feature extraction method. That is, CNN is used to extract packet spatial features, and RNN is used to further extract temporal features;
- This paper proposes an intrusion detection strategy based on GAN. In this strategy, the generator and discriminator of GAN are replaced by LSTM-RNN;
- An intrusion detection method based on adversarial domain adaptation is proposed, and performance tests are performed on three different datasets: ISCXIDS2012, CIC-IDS-2017 and NDSec-1.

2   Related Works

2.1   Research on Intrusion Detection Method Based on Deep Learning

With the continuous development of internet technology, the related technologies of malware and malicious attacks have become increasingly complex. Malicious attack techniques may use encryption, disguise, anonymity, and other methods to hide in normal data. Without effective identification methods, it is difficult to identify malicious data that has been invaded through network transmission. Compared to general machine learning, deep learning has great convenience in identifying malicious software.

Shibahara T et al. [2] proposed a method for malware identification that improves the classification ability of NRR by identifying two key features of malicious communication software and applying a recursive tensor network to compute the threat component of the input signature, the accuracy of the method is as high as 97.6%, but the method is limited by the sample labeling. O. E. David et al. [3] used DBN trained on unlabeled data to generate malware code, and then used DBN and autoencoder (AE) to identify and classify malware. Using traditional methods to generate malware signatures is easy to detect due to malware code modification, David et al. obtained logs through a sandbox, used n-gram model to process the logs, and then obtained the most common 20,000 unigram models that only appear in malware, and created a 20,000-dimensional feature vector to identify whether a given unigram model appeared. Then, these recognized unigram models are used to pre-train an 8-layer DBN with DAE. The final malware signature is a vector of 30 numbers, and experiments show that the trained model is 98.6% accurate.

2.2 Research on Adversarial Network Algorithm Based on GAN

Although machine learning technology has gradually become a mainstream network intrusion detection solution, due to the inherent security fragility of the model, it is difficult to resist adversarial attacks. Researchers have attempted to generate adversarial network traffic using different methods from the traffic dimension to avoid the identification of intrusion detection systems, and have achieved significant results.

The GAN-based generation method is mainly suitable for black-box attacks. The generator generates adversarial network traffic by modifying certain characteristics in the traffic, and the discriminator simulates a black-box model to assist training. At present, many GAN-based derivative frameworks have been proposed and applied in practical testing. Pan Yiming et al. [4] proposed a new framework for generating adversarial samples based on GAN, which consists of a generation module, an identification module, and a verification module. Unlike traditional GAN, which only use random noise as the generator input, the generator input of the new framework is a processed malicious traffic sample used to solve the problem that the generator cannot converge during optimization.

2.3 Research on Algorithms Based on Transfer Learning

We call the areas we have studied "Source domains", "Target domain" refers to the field we want to learn, and use T_S and T_T to represent the source task and the target task. Sinno Jialin Pan and Qiang Yang [5] divided transfer learning into three categories based on the similarity of the source and target domains, they are Inductive-Based Transfer Learning [6], Unsupervised Transfer Learning [7, 8] and Direct Push Transfer Learning [9].

Wenyuan Dai et al. [10] proposes an instance-based TrAdaBoost transfer learning algorithm based on the Boosting method: when the target sample is classified incorrectly, it is more difficult to classify the sample by default, and the weight of the sample is further increased, otherwise, the weight should be reduced. Sinno Jialin Pan et al. [11] proposed the TCA method for migration component analysis. This method first calculates the L and H matrices, then uses the commonly used kernel functions to map and calculate the feature values, and finally uses traditional machine learning methods to process the dimensionality reduction data. Busto P P and Gall J. [12] proposed OpenSet Domain Adaptation. This method uses the relationship between the source domain and the target domain to label the target domain sample and transform the source domain into the same space as the target domain, so that the learning label and the learning map alternate until convergence or the target value is less than a specific value.

3 IDS Based on Adversarial Domain Adaptation Algorithm

In this chapter, this paper proposes an intrusion detection method based on adversarial domain adaptation, first using CNN to extract the spatial features of packets, and then using the LSTM network to further extract temporal features and using anomaly detection algorithms based on discriminator loss and residual loss, and then using adversarial-based non-generative model methods to reduce the difference distance between the target domain and the source domain feature mapping space, and finally extracting the invariant features between the source domain and the target domain. This method only requires transfer learning at the data level, so it has good generalization. The method architecture proposed in this chapter is as follows (Fig. 1):

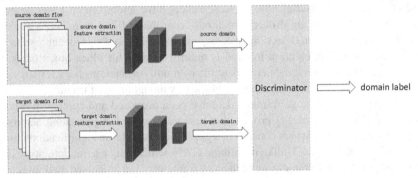

Fig. 1. Intrusion detection framework diagram based on adversarial domain adaptation algorithm.

3.1 Spatiotemporal Feature Extraction Module Based on CNN and RNN

Network traffic is hierarchical, with data typically consisting of packets, which in turn consist of multiple bytes. This article uses FlowMining to process flow files to generate a three-dimensional vector representing flow features, and performs zero padding around the original matrix to complete packet truncation and completion operations. Neural networks can improve data features at different levels, therefore, this paper draws inspiration from the way HAST-NAD uses CNN to process network packet spatial features and RNN to process network flow temporal features, uses anomaly detection based on generative adversarial training to handle spatiotemporal traffic. The framework is shown in the following Fig. 2.

3.2 Anomaly Detection Module Based on Discriminator Loss and Residual Loss

In this paper, we use a GAN-trained discriminator to learn an unsupervised way to perform anomaly detection [13]. In paper [14, 15], they updated the mapping from real-time space to specific latent space to enhance the training of generators and discriminators, in recent experiments, the researchers proposed to train latent spatial comprehensible GAN and applied it to unsupervised learning. By utilizing reconstructed test samples

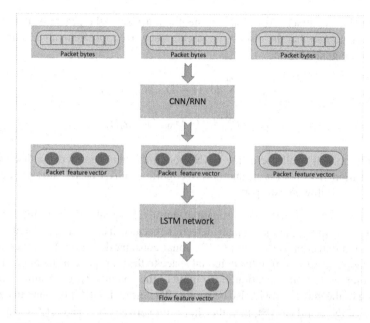

Fig. 2. Feature extraction framework diagram based on RNN and LSTM.

from latent space for anomaly identification and successfully applying the proposed GAN-based detection strategy, [16, 17] discover unexpected labeling of images. In this paper, based on these studies, we utilize GAN's residual loss and discriminant loss for anomaly detection.

LSTM-RNN have been shown to learn complex time series by acquiring information from memory cells backward (or even forward) time steps, and in this paper, in order to process the spatiotemporal data extracted by features, both the generator (G) and the discriminator (D) of GAN will be replaced by LSTM-RNN.

$$\min_{G} \max_{D} V(D, G) = \mathcal{E}_{x \sim p_{\text{data}}(x)}\left[logD(x)\right] + \mathcal{E}_{z \sim p_z(Z)}\left[\log(1 - D(G(z)))\right] \quad (1)$$

where, the generator G implicitly defines the probability distribution of the generated sample, which can be written as $G_{rnn}(z)$, z comes from the distribution of random latent space. Discriminator D is trained to minimize the average negative cross entropy between its prediction and sequence tags (e.g. training D to identify as many training samples as possible and identifying as few generated samples as possible). Therefore, the loss of the discriminator is shown in formula (2):

$$\begin{aligned} D_{\text{loss}} &= \frac{1}{m}\sum_{i=1}^{m}\left[logD_{rnn}(x_i) + log(1 - D_{rnn}(G_{rnn}(z_i)))\right] \\ &\Leftrightarrow \min\frac{1}{m}\sum_{i=1}^{m}\left[-logD_{rnn}(x_i) - log(1 - D_{rnn}(G_{rnn}(z_i)))\right] \end{aligned} \quad (2)$$

where x_i, $i = 1, \ldots, m$ represents the real training sample that should be recognized as a normal real sample, $G_{rnn}(z_i)$, $i = 1, \ldots, m$ is the dummy generated sample that should be identified as an attack. At the same time, the purpose of the generator is to confuse the discriminator so that the discriminator can recognize more generated samples, and the loss of the generator is:

$$G_{loss} = \sum_{i=1}^{m} log(1 - D_{rnn}(G_{rnn}(z_i)))$$
$$\Leftrightarrow min \sum_{i=1}^{m} log(-D_{rnn}(G_{rnn}(z_i))) \tag{3}$$

In order to make full use of the GAN model, the generator and the discriminator should be trained to help with anomaly recognition, GAN-based anomaly detection consists of the following two parts:

(1) Anomaly detection with discrimination: intuitively speaking, the trained discriminator D is a direct tool for anomaly detection, because it can distinguish data streams with high detection sensitivity; (2) Residual constant detection: the generator G is the mapping process from the potential space to the real space, as mentioned in [18] the potential smooth transition, if the input of the potential space is similar, then the output of the generator will also be similar. Therefore, if $G(Z^k)$ is found in the latent space of the test data X^{tes}, then the similarity between X^{tes} and $G(Z^k)$ can explain the extent to which X^{tes} follows the distribution reflected by G.

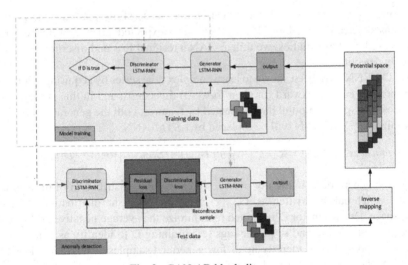

Fig. 3. GAN-AD block diagram.

As shown in the lower half of Fig. 3, in order to find the optimal Z^k corresponding to the test sample. The reconstructed original sample $G(Z^1)$ is first sampled from the latent space random set Z^1 and fed back to the generator, and then the space of the potential sample can be updated with the gradient obtained by the error function defined by X^{tes}

and G(Z). For simplicity, similarity between sequences can be defined as covariance.

$$\min_{Z^k} Er\left(X^{tes}, G_{rnn}\left(Z^k\right)\right) = 1 - Simi\left(X^{tes}, G_{rnn}\left(Z^k\right)\right) \tag{4}$$

If, after enough iteration loops, the error is small enough, sample Z^k is recorded as a corresponding map in the potential space of the test sample. So the residuals of the test sample at time t are calculated as follows:

$$Res\left(X_t^{tes}\right) = \sum_{i=1}^{n}\left|x_t^{tes,i} - G_{rnn}\left(Z_t^{k,i}\right)\right| \tag{5}$$

where $X_t^{tes} \subseteq \mathcal{R}^n$ is the measurement of n variables at time step t, and then the anomaly score for anomaly detection is calculated, and the formula is calculated:

$$S_t^{tes} = \lambda Res\left(X_t^{tes}\right) + (1 - \lambda)D_{rnn}\left(X_t^{tes}\right) \tag{6}$$

In this paper, the anomaly detection problem for multivariate time series is formulated as follows, first considering the m-dimensional time series $X = \left\{x^{(t)}, t = 1, \ldots, T\right\}$ with length T, where $x^{(t)} \in \mathcal{R}^m$ is the m-dimensional vector of m variables t at time point. The GAN model is trained on the normal time series dataset X^{real} and generates a fake sample X^{gs} that "looks real", next, test the time series dataset X^{att}. The trained model can be analyzed to detect anomalies, but the high-dimensional input of LSTM-RNN can result in higher computational costs than typical deep neural networks. Therefore, in this paper, we use principal component analysis (PCA) to project high-dimensional data into PC projection space, and then input the data into GAN model $X^{tes} \subseteq \mathcal{R}^m \Rightarrow \mathcal{X}^{tes} \subseteq \mathcal{R}^n$, projected as:

$$\begin{aligned} P &= PCA\left(X^{real}\right) \\ \mathcal{X}^{tes} &= X^{tes} P^T \end{aligned} \tag{7}$$

where, $X^{tes} \subseteq \mathcal{R}^m, \mathcal{X}^{tes} \subseteq \mathcal{R}^n, P \subseteq \mathcal{R}^{n \times n}$, m is the original dimension and n is the number of retained principal components. Then, the predictors are fed into the GAN-AD model. According to formula (6), the anomaly score is applied, and the label assignment function is applied to identify whether the i-th variable of the time series set $\mathcal{X}^{tes}$ tested at time step t is attacked.

$$A_t^{tes,i} = \begin{cases} 1, & \text{if } H\left(S\left(x_t^{tes,i}\right), 1\right) > \tau \\ 0, & \text{else} \end{cases} \tag{8}$$

where, $t = 1, \ldots, T, i = 1, \ldots, n$, if the spread entropy error $H(.,.)$ of the anomaly score is higher than a certain value τ, an anomaly is detected.

3.3 Intrusion Detection Algorithm Based on Adversarial Domain Adaptation

Deep Domain Adaptation adds the concept of domain adaptation in the process of transfer learning, which has more advantages than other transfer learning methods in mining transferable features, and the deep domain adaptation process is shown in Fig. 4 In this paper, D_S is used to represent the source domain, D_t represents the target domain, X represents the feature space, d represents the dimension of X, P(X) represents the edge distribution of X, T represents the task, T_S represents the task on the source domain, and T_t represents the task on the target domain.

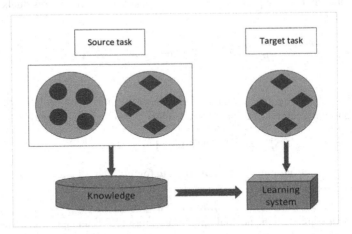

Fig. 4. Domain adaptation structure diagram.

This paper first uses CNN to extract the spatial features of the packet and obtain a two-dimensional vector, the two-dimensional vector can be regarded as a grayscale map, so the flow is a large picture composed of multiple small graphs, here can be borrowed from the feature extraction method of the picture, so this paper is based on the excellent effect of ADDA using adversarial domain in the image field to improve it on the basis of making it more suitable for anomaly detection.

As shown in Fig. 5, the model used in this paper contains three parts: source domain feature extractor E_S, target domain feature extractor E_t, and discriminator D_d.

The target domain feature extractor E_t is equivalent to the usual generator in GAN, and it is isomorphic to the source domain feature extractor E_S, this paper uses the feature extractor cascaded by CNN and RNN which has been trained above to extract the features of the network data stream, then E_t will initialize the parameters that have been trained to speed up the convergence speed of the training, keep the parameters of E_s unchanged during the training process, and constantly adjust E_t. After many adversarial training, the target domain feature extractor can extract potential features similar to those extracted by the source domain feature extractor. Therefore, it is only necessary to train classifier C_s in the source domain, and then the features extracted by the target domain feature extractor E_t target domain D_t can also be well classified, in which the domain migration is completed by potential feature mapping, the potential feature mapping is carried out through the feature extractor, and the feature extractor is optimized by adversarial training.

In the process of training, the parameters of the discriminator D_d are constantly updated, and then the discriminator is deceived by adversarial training. The source data

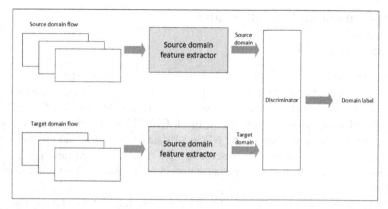

Fig. 5. Frame Diagram of intrusion Detection algorithm based on adversarial Domain.

set of E_s input includes normal flow and abnormal traffic, and unlabeled data is input to E_t. Finally, E_s and E_t input the extracted spatio-temporal features into the discriminator to make the discriminator continuously improve its discrimination ability. After many confrontations, E_t can output features similar to E_s.

Starting from the actual application scenario, this paper fully takes into account the changes of data flow characteristics of different networks in the actual scene and the progress of network attack and defense, and the problem that offline data is difficult to be directly applied to the actual scene. An intrusion detection algorithm based on adversarial domain adaptation is proposed.

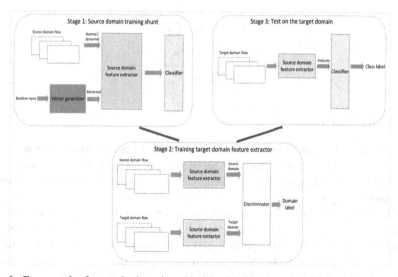

Fig. 6. Framework of anomaly detection algorithm based on adversarial domain adaptation.

4 Experiment and Result Analysis

4.1 Experimental Setup

The experimental scenario in this article requires direct anomaly detection of network traffic. Therefore, after extensive research, ISCXISDS2012, NDSec-1, and CIC-ISD-2017 were selected as the validation datasets for this article based on factors such as data size and attack type. Among them, the ISCX2012 dataset is currently a widely used intrusion detection dataset. The NDSec-1 dataset contains tracking and log files of network attacks synthesized by researchers from network facilities. The CIC-ISD-2017 dataset contains benign and the latest common attacks, similar to real world data (PCAPs).

This paper uses accuracy rate (ACC), false positive rate (FAR), and anomaly detection rate (DR) to evaluate model performance. The F1 score is the weighted and average of accuracy and recall, and TP is used to represent the number of samples that were actually correctly identified as abnormal traffic samples. TN represents the correct number of samples that were actually identified as normal, FP represents the number of samples that were misidentified as normal flowing samples, and FN represents the number of samples that were actually misidentified as normal samples. The calculation formulas are as follows:

$$ACC = \frac{TP + TN}{TP + FP + FN + TN} \tag{9}$$

$$DR = \frac{TP}{TP + FN} \tag{10}$$

$$FAR = \frac{FP}{FP + TN} \tag{11}$$

$$F1 = \frac{2TP}{2TP + FN + FP} \tag{12}$$

4.2 Results and Analysis

Anomaly Detection and Verification of Generative Confrontation Training

As shown in Fig. 7, this paper uses spatiotemporal feature extraction based on CNN and RNN to extract features from the original data stream, and then uses the anomaly detection of generative adversarial training to calculate the accuracy, F1 score and detection rate of three different datasets: where the accuracy of the NDSec-1 dataset and the F1 score, the anomaly detection rates were 0.9976, 0.9953, 0.9951, the accuracy and F1 score of the ISCXIDS2012 dataset, and the anomaly detection rate were 0.9958, 0.9922, 0.9914, and the accuracy of the CIC-IDS-2017 dataset, the F1 score and the anomaly detection rate were 0.9948, 0.9955 and 0.9985, respectively. It can be seen from the figure that the NDSec-1 dataset has the best detection effect, and the accuracy of ISCXIDS2012 and CIC-IDS-2017 dataset is relatively poor, but the detection rate level is relatively high.

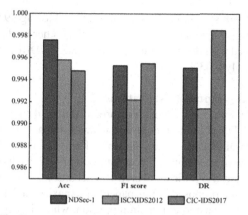

Fig. 7. Performance comparison on different datasets.

With the wide application of deep learning algorithms, different machine learning algorithms have gradually become excessive to deep learning, and anomaly detection based on deep learning can train classifiers with better effect without feature engineering, so this paper also uses other deep learning algorithms to compare with the methods proposed in this paper. Since the CIC-IDS-2017 dataset attack type is more comprehensive and has the most natural background traffic, three different anomaly detection algorithms based on the CIC-IDS-2017 dataset are selected for comparison, namely LUCID [19], STDeepGraph [20], UDBB [21], and the comparison results are shown in the following table:

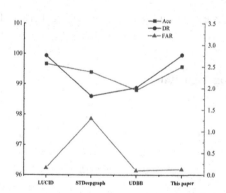

Fig. 8. Comparison of experiments on the CIC-IDS-2017 dataset.

From Fig. 8, it can be concluded that on the CIC-IDS-2017 dataset, the three methods used in this paper are used to compare with the anomaly detection method proposed in this paper, and LUCID achieves the best results in terms of accuracy, with an accuracy rate of 99.67%, and the anomaly detection algorithm based on discriminator loss and residual loss proposed in this paper is only 0.09% worse than the LUCID algorithm in terms of accuracy. The anomaly detection algorithm proposed in this paper has the best performance in terms of anomaly detection rate, achieving 99.94% accuracy. The method proposed in this paper also performs better than other methods in terms of false positive rate, compared with the 0.1% false positive rate of UDBB, and the false positive rate in this paper is only 0.03%.

Adversarial Domain Adaptation Algorithms

In this paper, the model is first trained on three different datasets of ISCXIDS2012, CIC-IDS-2017 and NDSec-1, and then tested without domain adaptation algorithm on another dataset, the training results are shown in Fig. 9. Compared with the intrusion detection efficiency comparison on different datasets in Fig. 7, the model without domain adaptation algorithm has a significant decrease in accuracy, F1 score and detection rate. For example, the method proposed in this paper has a maximum detection rate of 79.6% and a minimum of 69.5% without domain migration, which is significantly lower than that of the test results on the source data.

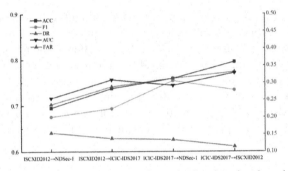

Fig. 9. The model migration effect is not performed using domain adaptation algorithms.

Use Transfer Learning Algorithms

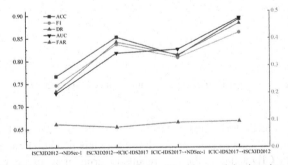

Fig. 10. Use the domain adaptation algorithm for model migration effects.

As shown in Fig. 10, if different data sets use domain adaptation algorithms for migration, their detection accuracy, F1 index, false alarm rate, etc. have been significantly improved, although there are still some gaps compared to Fig. 6, but this is a normal situation, because the proportion of abnormal data in different data sets, the types of abnormal data are also different, this paper uses an adversarial domain adaptation algorithm based on unsupervised learning, because no label information is used, so the effect is still a certain gap compared to supervised learning. After using the domain adaptation algorithm, it can be seen that the accuracy index is the highest 89.4%, which is 8% higher than before, the lowest is 76.5%, and the increase is 7% compared with the unused domain adaptation algorithm, which also shows that the method proposed in this paper is effective.

Comparison with Other Methods
At present, the application of transfer learning in network anomaly detection is relatively small, so for the ISCXIDS2012, CIC-IDS-2017 and NDSec-1 datasets in this paper, the two papers [22] and [23] are reproduced on this basis, of which the paper [22] method is the classical transfer learning method and [23] is the latest transfer learning method, and the two methods are trained and tested by the data set processed in this paper, and the final results are compared with the results of this paper The experimental results are shown in Fig. 11:

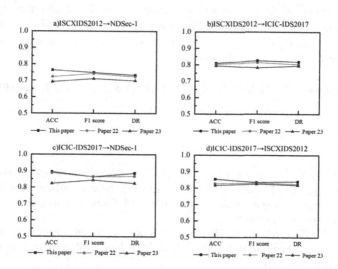

Fig. 11. Comparison of different transfer learning methods.

From the results of the above figure, it can be concluded that compared with the method proposed in the paper [22, 23], the accuracy rate of the method proposed in this paper is 89.3%, which is 1.1% and 7.0% higher than the other two methods, and the F1 score is 86.3%, which is 0.2% and 1.9% higher than the other two methods, respectively, and the detection rate is 88.6%, and the detection rate is 1.8% and 6.0% higher than the other two methods, respectively. In the dataset ISCXID2012 to dataset IDS2017, dataset

IDS2017 to dataset NDSec-1, dataset IDS2017 to dataset ISCXID2012, the method proposed in this paper has improved accuracy, detection rate and F1 index compared with other methods.

5 Summary and Outlook

Although the development of the Internet has brought convenience to people's lives in all aspects, with people's dependence on the Internet, criminals have discovered opportunities to make profits. Although most people's network security literacy has improved, they are still struggling with hacker intrusion methods in the face of professional equipment and knowledge. As the second line of defense for network security, intrusion detection mainly collects, monitors, analyzes data, and alerts potential intrusion detection behaviors, which can effectively prevent the vast majority of conventional attacks.

Therefore, this article proposes an intrusion detection system based on adversarial domain adaptation algorithms. Using CNN and RNN based spatiotemporal feature extraction algorithms to extract different levels of features from the flow, the discriminator and classifier of GAN are replaced with long and short term recurrent neural networks. Through multiple iterations, the discriminator loss and residual loss are used for anomaly detection. Then, this article uses adversarial domain algorithms to map the target domain and source domain to the same feature space, and after multiple rounds of adversarial training, the goal of domain migration is achieved. However, compared with other algorithms, the method proposed in this paper is not fully leading, so there is still a lot of optimization space for the method proposed in this paper, and transfer learning related content still needs to be focused on.

Acknowledgement. This work was supported by the Key R&D Program of Jiangsu (BE2022081).

References

1. Lunt, T.F.: A survey of intrusion detection techniques. Comput. Secur. **12**(4), 405–418 (1993)
2. Shibahara, T., Yagi, T., Akiyama, M., et al.: Efficient dynamic malware analysis based on network behavior using deep learning. In: 2016 IEEE Global Communications Conference (GLOBECOM), pp. 1–7. IEEE (2016)
3. David, O.E., Netanyahu, N.S.: DeepSign: deep learning for automatic malware signature generation and classification. In: 2015 International Joint Conference on Neural Networks (IJCNN), pp. 1–8. IEEE (2015)
4. Pan, Y., Lin, J.: Malicious network stream generation and verification based on generative adversarial networks. J. East China Univ. Sci. Technol. **45**(2), 165–171 (2019)
5. Pan, S.J., Yang, Q.: A survey on transfer learning. IEEE Trans. Knowl. Discov. Data Eng. **22**(10), 1345–1359 (2010)
6. Fengmei, W., Jianpei, Z., Yan, C., et al.: FSFP: transfer learning from long texts to the short. Appl. Math. Inf. Sci. **8**(4), 2033 (2014)
7. Dai, W., Yang, Q., Xue, G.R., et al.: Self-taught clustering. In: Proceedings of the 25th International Conference on Machine Learning, pp. 200–207 (2008)

8. Samanta, S., Selvan, A.T., Das, S.: Cross-domain clustering performed by transfer of knowledge across domains. In: 2013 Fourth National Conference on Computer Vision, Pattern Recognition, Image Processing and Graphics (NCVPRIPG), pp. 1–4. IEEE (2013)
9. Dai, W., Xue, G.R., Yang, Q., et al.: Co-clustering based classification for out-of-domain documents. In: Proceedings of the 13th ACM SIGKDD International Conference on Knowledge Discovery and Data Mining, pp. 210–219 (2007)
10. Dai, W.Y., Yang, Q., Xue, G.R., et al.: Boosting for transfer learning. In: Proceedings of the 24th International Conference on Machine Learning, pp. 193–200. Morgan Kaufmann Publishers, San Francisco (2007)
11. Pan, S.J., Tsang, I.W., Kwok, J.T., Yang, Q.: Domain adaptation via transfer component analysis. IEEE Trans. Neural Networks 22(2), 199–210 (2011). https://doi.org/10.1109/TNN.2010.2091281
12. Busto, P.P., Gall, J.: Open set domain adaptation. In: Proceedings of 2017 IEEE International Conference on Computer Vision, pp. 754–763. IEEE, Venice (2017)
13. Xue, Y., Xu, T., Zhang, H., et al.: SegAN: adversarial network with multi-scale L1 loss for medical image segmentation. Neuroinformatics 16(3), 383–392 (2018)
14. Yeh, R.A., Chen, C., Yian Lim, T., et al.: Semantic image inpainting with deep generative models. In: Proceedings of the IEEE Conference on Computer Vision and Pattern Recognition, pp. 5485–5493 (2017)
15. Salimans, T., Goodfellow, I., Zaremba, W., et al.: Improved techniques for training GANs. In: Advances in Neural Information Processing Systems, vol. 29 (2016)
16. Schlegl, T., Seeböck, P., Waldstein, S.M., Schmidt-Erfurth, U., Langs, G.: Unsupervised anomaly detection with generative adversarial networks to guide marker discovery. In: Niethammer, M., et al. (eds.) IPMI 2017. LNCS, vol. 10265, pp. 146–157. Springer, Cham (2017). https://doi.org/10.1007/978-3-319-59050-9_12
17. Zenati, H., Foo, C.S., Lecouat, B., et al.: Efficient GAN-based anomaly detection. arXiv preprint arXiv:1802.06222 (2018)
18. Radford, A., Metz, L., Chintala, S.: Unsupervised representation learning with deep convolutional generative adversarial networks. arXiv preprint arXiv:1511.06434 (2015)
19. Doriguzzi-Corin, R., Millar, S., Scott-Hayward, S., et al.: LUCID: a practical, lightweight deep learning solution for DDoS attack detection. IEEE Trans. Netw. Serv. Manage. 17(2), 876–889 (2020)
20. Yao, Y., Su, L., Lu, Z., et al.: STDeepGraph: spatial-temporal deep learning on communication graphs for long-term network attack detection. In: 2019 18th IEEE International Conference on Trust, Security and Privacy in Computing and Communications/13th IEEE International Conference on Big Data Science and Engineering (TrustCom/BigDataSE), pp. 120–127. IEEE (2019)
21. Abdulhammed, R., Musafer, H., Alessa, A., et al.: Features dimensionality reduction approaches for machine learning based network intrusion detection. Electronics 8(3), 322 (2019)
22. Ghifary, M., Kleijn, W.B., Zhang, M., Balduzzi, D., Li, W.: Deep reconstruction-classification networks for unsupervised domain adaptation. In: Leibe, B., Matas, J., Sebe, N., Welling, M. (eds.) ECCV 2016. LNCS, vol. 9908, pp. 597–613. Springer, Cham (2016). https://doi.org/10.1007/978-3-319-46493-0_36
23. Niu, J., Zhang, Y., Liu, D., et al.: Abnormal network traffic detection based on transfer component analysis. In: 2019 IEEE International Conference on Communications Workshops (ICC Workshops), pp. 1–6. IEEE (2019)

Small-Sample Coal-Rock Recognition Model Based on MFSC and Siamese Neural Network

Guangshuo Li[1], Lingling Cui[2], Yue Song[1], Xiaoxia Chen[3], and Lingxiao Zheng[1(✉)]

[1] Shandong University of Science and Technology, Qingdao 266590, Shandong, China
350286119@qq.com
[2] Foundation Department, Shandong Water Conservancy Vocational College, Rizhao 276826, Shandong, China
[3] Zaozhuang Mining Group Coal Washing and Dressing Center, Zaozhuang 277000, Shandong, China

Abstract. Given the advantages of deep learning in feature extraction and learning ability, it has been used in coal-rock recognition. Deep learning techniques rely on a large number of independent identically distributed samples. However, the complexity and variability of coal-rock deposit states make the dataset exhibit small sample characteristics, resulting in poor performance of deep learning model. To address this problem, this paper proposes a framework named MFSC-Siamese, which combines the advantages of log Mel-Frequency Spectral Coefficients (MFSC) and Siamese neural network. First, the MFSC is used to extract vibration signal features to preserve the information of the original signal as much as possible, which makes the extraction of vibration features more accurate. Second, a recognition model based on Siamese neural network is proposed to reduce the number of participants by sharing network branches, which achieves coal-rock recognition by learning the distance between sample features, closing the distance between similar samples and distancing the distance between dissimilar samples. To evaluate the effectiveness of the proposed method, a real vibration signal dataset was used for comparative experiments. The experimental results show that the proposed method has better generalization performance and efficiency, with accuracy up to 98.41%, which is of great significance for the construction of intelligent mines.

Keywords: Coal-rock recognition · Small-Sample · MFSC · Siamese Neural Network

1 Introduction

The traditional manual coal mining method is difficult to accurately recognize coal-rock interface, which easily leads to under-cutting and over-cutting of the coal seam, causing great difficulties in the efficient production and utilization of coal [1]. With the promotion of intelligent construction of coal mines, efficient coal-rock recognition has become a challenge that needs to be solved for intelligent coal mining.

The vibration signal of mining machinery during coal mining is a relatively stable and reliable source of information [2], and therefore has been applied to coal-rock recognition. Liu [3] divided the extracted vibration signal into three characteristic frequency bands (low, medium, and high) and performed coal-rock recognition based on the energy of the characteristic frequency bands. Liu et al. [4] constructed a vibration model for the mining. Zhao et al. [5] developed a coal-rock cutting state recognition system based on fuzzy control by utilizing the vibration characteristics of the shearer drum and rocker arm. Zhang et al. [6] proposed a coal-rock cutting state recognition method for the shearer based on the back-coupling distance analysis of vibration signals. Liu et al. [7] used wavelet packet energy method to convert the vibration signal from mode space to feature space and proposed a recognition model based on wavelet packet decomposition and fuzzy neural network. Although the existing methods have made some progress in coal-rock recognition based on vibration signals, however, these methods are difficult to cope with the complex and variable nature of coal-rock endowment and mining conditions. Specifically, the coal-rock cutting vibration signal can vary widely even within the same mining site, due to complex factors such as coal-rock type, coal-rock percentage, burial depth, and mining machine type. Therefore, it is undoubtedly difficult or even impractical to collect a large number of independent and identically distributed (i.i.d.) samples, which makes the traditional neural network models underperform due to the unavailability of sufficient training data.

MFSC is a widely used signal feature extraction method that can adequately capture the features of a signal while reducing the number of samples required to train a deep neural network [8]. Siamese neural network is a class of network architectures that usually contains two identical subnetworks. The twin subnetworks have the same configuration with the same parameters and shared weights. This architecture with shared weights reduces the number of parameters and can be better applied to small-sample dataset [9]. Therefore, we propose a coal-rock recognition model based on MFSC and Siamese neural network (MFSC-Siamese) for small-sample coal-rock cutting vibration signals. The main contributions of this paper are as follows:

1) MFSC is used to extract features of coal-rock cutting vibration signals to improve the accuracy of coal-rock recognition by reducing the information loss of feature extraction.
2) Siamese neural network is applied to coal-rock recognition for the first time to improve the generalization ability and robustness of the model. The Siamese neural network alleviates the overfitting problem easily brought by using small samples to train large models, and greatly reduces the number of model parameters by sharing weights.
3) The experiments constructed by real dataset reflects the superiority of the proposed method and makes a new attempt for coal-rock recognition.

2 Proposed Method

In this paper, an effective method is proposed for coal-rock recognition by coal-rock cutting vibration signals. The method improves the accuracy and generalization ability of small-sample coal-rock recognition, maximizes the feature extraction of vibration signals by MFSC, and reduces the number of parameters by Siamese neural networks. As shown in Fig. 1, the overall architecture of the model mainly consists of two parts.

First, MFCS decomposes the two vibration signals to generate the MFSC features, which are used as the inputs to the twin networks. Then, the spectro-temporal feature vectors are extracted from the branch of the weight-shared convolutional neural network, and the Euclidean distance between the feature vectors is used as the parameter of the loss optimization model.

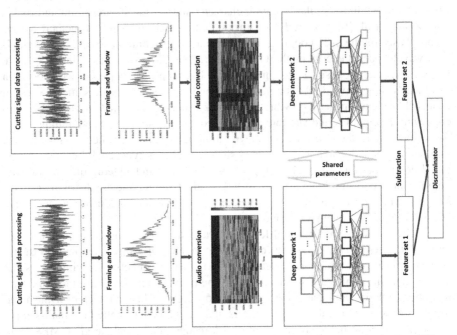

Fig. 1. The overall architecture of our proposed model. Note that "Deep Network 1" and "Deep Network 2" are weight-shared.

2.1 Mel Spectrum Feature Extraction

The proposed method uses MFSC as spectral features of vibration signal. MFSC uses a nonlinear Mel filter to filter the Fourier transformed signal to obtain the spectral features of audio signal. The reason for using MFSC instead of the more common MFCC is that, MFCC has an additional DCT than MFSC, which not only requires longer processing time but also causes serious loss of high frequency nonlinear components in the original vibration signal. The DCT is a linear transformation that can be completely replaced by the neural network itself [8].

Pre-emphasis
Pre-emphasis amplifies the high frequency part of signal. After the pre-emphasis filter, the noise is suppressed. The pre-emphasis filter uses a first-order filter as follows:

$$y(t) = x(t) - \alpha x(t - 1) \tag{1}$$

where x is the original signal, y is the filtered signal, and the filter coefficient value α is 0.97, which is the default value commonly used for generating the Mel spectrum.

Framing and Window

After applying pre-emphasis to the signal, it needs to be segmented into shorter frames in order to capture the changes in frequency over time. Specifically, we used a frame size of 25 ms and a stride of 10 ms (with an overlap of 15 ms) to ensure smooth transitions between adjacent frames. Then, we apply the Hamming window to each frame. Hamming Windows have the following forms:

$$w[n] = 0.54 - 0.46 \cos\left(\frac{2\pi n}{N - 1}\right) \tag{2}$$

where $0 \le n \le N - 10$, N is the window length. Compared with the ordinary rectangular window function, the spectral leakage can be reduced by Hamming window [10, 11]. After framing and window, the signal is shown in Fig. 2(a).

Fast Fourier-Transform and Power Spectrum

Now the time-domain signal is converted into the frequency domain via an N-point FFT for each frame. The power spectrum (periodogram) is calculated using the following formula:

$$P = \frac{|FFT(x_i)^2|}{N} \tag{3}$$

where N is set to 256, x_i is the i-th frame of signal x.

Mel Frequency Filters

The amplitudes in the frequency domain are refined using a filter bank. At each frequency, the product of periodogram P(k) and filter Hm(k) is calculated is calculated to define a triangular filter bank with 40 filters. The final Mel spectrum is obtained as shown in Fig. 2(b) by filtering the power spectrum applied to the signal.

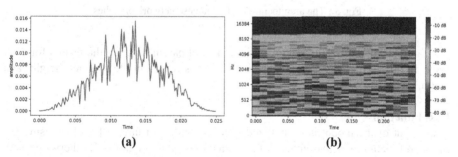

(a)   (b)

Fig. 2. The phased results of extracting Mel features, where (a) is the signal after framing and windowing, and (b) is the Mel spectrum of the signal.

Mask

Time-frequency (T-F) masking is an effective method to address model overfitting by means of data augmentation [12]. We take the average of 2 regions of width 0.1 in the time dimension as the masks, and the same for the frequency dimension.

2.2 Network Architecture and Training

The Siamese neural network is used to construct a coal-rock recognition model with two subnetworks having the same architecture, parameters and weights. The two subnetworks extract features with different inputs, and use the Euclidean distance to measure the differences between features.

The Architecture of Our Siamese Network

As shown in Fig. 3, the convolutional neural network is used to construct twin network branches in this paper.

Each of the two network branches takes a Mel spectrum of size 256×256 as input. A tensor of 32 @ 28×28 is obtained by 4 convolutional layers (Conv) with a kernel size of 3×3. Finally, the dimensionality of the tensor is converted to 25088 using a fully connected layer (FC).

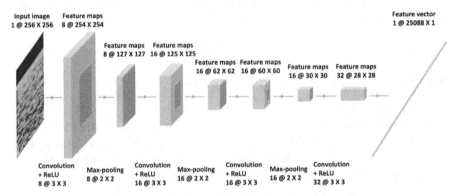

Fig. 3. The architecture of the Siamese network branches

Figure 3 illustrates all the convolutional kernel parameters and the results for each layer. In this format, C @ H×W, these parameters represent the channel, height, and width, respectively.

Loss Function

The output of the model is normalized to a probability of having [0, 1] by using the Sigmoid function. For two images of the same class, the output t is 1, otherwise t is 0. Logistic regression is used for training, which means that the loss function should be a binary cross-entropy between the prediction and the target. There is also a L2 weight decay term in the loss function to allow the network to learn smaller and smoother weights to improve model generalization:

$$L(x_1, x_2, t) = t \cdot \log(p(x_1 \circ x_2)) + (1 - t) \cdot \log(1 - p(x_1 \circ x_2)) + \lambda \cdot \|w\|_2 \quad (4)$$

where $p(x_1 \circ x_2)$ denotes the probability that x_1 and x_2 belong to the same class.

Learning Schedule

In this paper, the model is optimized using the Adam algorithm. Specifically, the Adam optimizer uses the mean and variance of the gradients to estimate the first- and second-order moments for each parameter:

$$
\begin{cases}
m_t = \beta_1 m_{t-1} + (1 - \beta_1) g_t \\
v_t = \beta_2 v_{t-1} + (1 - \beta_2) g_t^2
\end{cases}
\tag{5}
$$

where g_t denotes the gradient of the t-th epoch, m_t and v_t denote the first- and second-order moments of the t-th epoch, $\beta_1 = 0.9$ and $\beta_2 = 0.999$ are the decay factors.

As m_t and v_t are small in the initial phase, bias correction is required. Specifically, we define $\hat{m}_t$ and $\hat{v}_t$ as the bias corrections of m_t and v_t:

$$
\begin{cases}
\hat{m}_t = m_t / 1 - \beta_1^t \\
\hat{v}_t = v_t / 1 - \beta_2^t
\end{cases}
\tag{6}
$$

Eventually Adam updates the parameters using the following equation:

$$
\theta_{t+1} = \theta_t - \frac{\alpha}{\sqrt{\hat{v}_t} + \varepsilon} \hat{m}_t
\tag{7}
$$

where θ_t denotes the parameter of the t-th epoch, and $\alpha = 0.0001$ is the learning rate, and $\varepsilon = 1e - 08$ is a constant that prevents the divisor from being zero. By dynamically adjusting the learning rate of the parameters, Adam enables the model to converge faster and more stably.

Effective Sample Construction

First, it needs to be clear that the input to the Siamese neural network is pairs of spectral features rather than a single one. With this type of sample construction, there will be a quadratic order of magnitude of spectral feature pairs used to train a model, which makes it difficult to overfit. Assuming a dataset with N classes and E samples per class, the total number of valid samples is:

$$
N_{pairs} = \binom{N \cdot E}{2} = \frac{(N \cdot E)!}{2!(N \cdot E - 2)!}
\tag{8}
$$

In addition, in order to avoid the issue of imbalanced samples, it is necessary to ensure that the pairwise combinations from the same and different classes are balanced. If each class has no less than E samples, there are $E!/2!(E - 2)!$ pairwise combinations per class and the number of pairwise combinations from the same category is $N_{same} = N \cdot E!/2!(E - 2)!$, where N_{same} grows exponentially as E increases and grows linearly as N increases.

3 Experiments

In this section, the dataset, experimental environment, parameter turning, results and discussions are presented.

3.1 Datasets

The dataset contains vibration signals generated by a MG180420-BWD coal-mining machine cutting coal and rock at a comprehensive mining face in Erdos, Inner Mongolia Autonomous Region. A magnetic suction-type wireless vibration sensor with a sampling frequency of 16 kHz is installed at the 5th axis of the left swing arm of the coal-mining machine. The data is cleaned and normalized to ensure the data quality.

The short-duration vibration signals are removed. The Mel spectrum features are extracted from the signals by MSFC after framing and windowing. Two classes of samples are obtained, with 280 vibration signals of cutting coal and 164 vibration signals of cutting rock. For class balancing, the samples of both classes are expanded to 500 using Mask. Finally, 1000 Mel spectrum features are divided into training set and test set in the ratio of 8:2.

3.2 Experimental Environment

The laboratory server is used to set up the experimental environment. The CPU is AMD Ryzen 5 4600H with Radeon Graphics@3.00GHz, the RAM is 16GB, the graphics card is NVIDIA GeForce GTX 1650 Ti, the operating system is Windows 10 (64-bit), the programming language is Python 3.8.13, and the deep learning framework is PyTorch 1.12 with CUDA 11.7.

3.3 Parameter Tuning

To select the optimal model parameters, we set the number of iterations to 500 and record the test accuracy during training. Figure 4(a) shows the accuracy of 500 iterations, with the blue line indicating the training accuracy and the orange line indicating the testing accuracy. The testing accuracy no longer improves significantly after 200 iterations and no longer fluctuates drastically. At this point, continued iteration will result in overfitting problems. We take the testing accuracy at 200 iteration as the accuracy of a single experiment. The final accuracy of our proposed model is 98.41% obtained by ten times of the same experiments.

As shown in Fig. 4(b), the AUC (Area under the Curve of ROC) is 0.9922, which is almost close to 1. This indicates that for this coal-rock cutting vibration signal dataset, our proposed MFSC-Siamese model arrives at a very high accuracy of coal-rock recognition. For the other evaluation metrics, the accuracy P is 98.41%, the recall R is 96.88%, and the F1 value is 97.64%.

3.4 Results and Discussions

To evaluate the effectiveness of our proposed Siamese neural network-based model, the features extracted by MFSC are used in existing models (1D-CNN, SVM, KNN) for comparative analysis of recognition results. To be fair, we designed the backbone of the 1D-CNN model to be consistent with our Siamese network branch, together with a fully connected layer and a sigmoid layer. In machine learning, too much dimensionality can lead to the "dimensional disaster" problem [13]. This is the problem that traditional

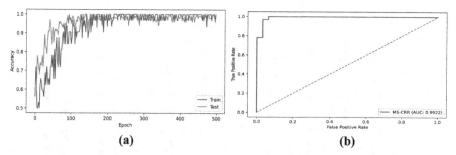

Fig. 4. Visualization of training process, where (a) is accuracy curve and (b) is ROC curve.

methods have when dealing with small samples. For this reason, we adopt PCA to reduce the dimensionality of the signal and determine the optimal feature dimension of 25 by cross-validation. For the SVM method, we use a Gaussian kernel as the kernel function and use three optimization methods to search for optimal parameters, namely GA (Genetic Algorithm), GS (Grid Search) and PSO (Particle Swarm Optimization).

To evaluate the performance of all models more accurately, ten experiments are conducted, and the experimental results of different models are shown in Table 1 and Fig. 5. Table 1 shows the average accuracy of six models.

Table 1. Model average accuracy

Method	Highest accuracy (%)	Lowest accuracy (%)	Mean (%)
1DCNN	97.31	93.62	95.32
GA-SVM	95.84	92.86	94.59
GS-SVM	94.88	92.56	93.65
PSO-SVM	93.99	91.44	93.12
KNN	85.76	81.36	83.43
Ours	99.54	97.06	98.41

The experimental results show that the average accuracy of our proposed method is 98.41%, which is 3.09% higher than the accuracy of 1D-CNN of the same architecture. Compared with SVM and KNN models, the performance of our model is improved by 3.82%, 4.76%, 5.29% and 14.98%. The coal-rock recognition method based on ordinary convolutional neural network and traditional machine learning shows poor performance. This is because that the 1D-CNN model is unable to learn the correct classification surface due to the small number of valid samples. Although dimensionality reduction by PCA enables SVM and KNN to avoid the "dimensional disaster" problem, it leads to some loss of original information.

In addition, efficiency is also an important factor to be considered in practical applications. Table 2 shows the average training time of the six models in ten training sessions.

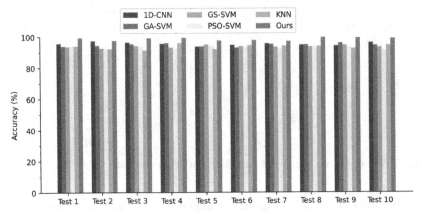

Fig. 5. The accuracy of all models.

Table 2. The comparison of training time

Method	Time(s)
1DCNN	47.15
GA-SVM	7905.89
GS-SVM	6114.65
PSO-SVM	10021.57
KNN	0.12
Ours	35.57

It can be seen that our method is much faster to train than the SVM method. Although KNN is faster, its accuracy is significantly lower. Compared with the 1D-CNN model, our model has some advantages in terms of efficiency.

4   Conclusion

In summary, a small-sample coal-rock recognition model (MFSC-Siamese) based on MFSC and Siamese neural network is proposed in this paper. The model achieves high accuracy and low-cost recognition of small-sample coal-rock cutting vibration signals, which is of great significance for future intelligent coal mining. The experimental results show that our method outperforms both traditional machine learning methods and ordinary convolutional neural network methods in terms of accuracy and efficiency, which provides a new idea for the development of intelligent technology in the coal mining field. In future work, we will explore the feasibility of this method for coal-rock recognition based on acoustic emission signals, as well as to explore potential improvements in its design and implementation.

References

1. Zhang, Q., Zhang, R., Liu, J., Wang, C., Zhang, H., Tian, Y.: Review on coal and rock identification technology for intelligent mining in coal mines. Coal Sci. Technol. **50**(2), 1–26 (2022)
2. Liu, C., Liu, Y., Liu, R., Bai, Y., Li, D., Shen, J.: Correlation load characteristic model between shearer cutting state and coal-rock recognition. J. China Coal Soc. **47**(1), 527–540 (2022)
3. Liu, J.: Study on Shearer Dynamic Seismic Response and Coal Rock Identification Technology. China University of Mining and Technology, Beijing (2020)
4. Liu, L., Zhao, H., Li, C.: Coal-rock recognition and control system of shearer based on vibration characteristics analysis. Coal Sci. Technol. **41**(10), 93–95 (2013)
5. Zhao, L., Wang, Y., Zhang, M., Jin, X., Liu, H.: Research on self-adaptive cutting control strategy of shearer in complex coal seam. J. China Coal Soc. **47**(1), 541–563 (2022)
6. Zhang, Q., Qiu, J., Zhuang, D.: Vibration signal identification of coal-rock cutting of shearer based on cepstral distance. Ind. Mine Autom. **43**(1), 9–12 (2017)
7. Liu, Y., Dhakal, S., Hao, B., Zhang, W.: Coal and rock interface identification based on wavelet packet decomposition and fuzzy neural network. J. Intell. Fuzzy Syst. **38**(4), 3949–3959 (2020)
8. Mohamed, A.-R.: Deep Neural Network Acoustic Models for ASR. University of Toronto Libraries, Toronto (2014)
9. Koch, G., Zemel, R., Salakhutdinov, R.: Siamese neural networks for one-shot image recognition. In: ICML Deep Learning Workshop, vol. 2, No. 1 (2015)
10. Astuti, W., Sediono, W., Aibinu, A.M., Akmeliawati, R., Salami, M.J.: Adaptive Short Time Fourier Transform (STFT) analysis of seismic electric signal (SES): a comparison of Hamming and rectangular window. In: 2012 IEEE Symposium on Industrial Electronics and Applications, pp. 372–377 (2012)
11. Trang, H., Loc, T.H., Nam, H.: Proposed combination of PCA and MFCC feature extraction in speech recognition system. In: 2014 International Conference on Advanced Technologies for Communications (ATC 2014), pp. 697–702 (2014)
12. Park, D.S., et al.: SpecAugment: a simple data augmentation method for automatic speech recognition. Proc. Interspeech **2019**, 2613–2617 (2019)
13. Wright, E.: Adaptive Control Processes: A Guided Tour. By Richard Bellman. 1961. 42s. Pp. xvi 255. (Princeton University Press). Math. Gazette **46**(356), 160–161 (1962)

Elemental Attention Mechanism-Guided Progressive Rain Removal Algorithm

Xingzhi Chen[1,2], Ruiqiang Ma[1,3(✉)], Shanjun Zhang[1,2,3], and Xiaokang Zhou[1,2,3]

[1] Inner Mongolia University of Technology, Hohhot, China
ruiq.ma@qq.com
[2] CISDI Information Technology (Chongqing) Co., Ltd., Chongqing, China
[3] Meiji University, Tokyo, Japan

Abstract. De-rainy has become a pre-processing task for most computer vision systems. Combining recursive ideas to De-rainy models is currently popular. In this paper, the EAPRN model is proposed by introducing the elemental attention mechanism in the progressive residual network model. The elemental attention mainly consists of spatial attention and channel attention, which feature-weight the feature image in both spatial and channel dimensions and combine as elemental attention features. The introduction of elemental attention can help the model improve its fitness for the rain removal task, filter out important network layers and help the network process the rainy image. Experiments show that the EAPRN model has better visual results on different datasets and the quality of the De-rainy image is further improved.

Keywords: Recursive · Attention Image · De-rainy · Residual Network

1 Introduction

Rainy weather affects people's travel in daily life, and it also affects outdoor work tasks. In rainy weather, rain usually exists in the form of rain streaks in images taken outdoors, as shown in Fig. 1. Rain streaks can cause image details to be covered or lost, while in heavy rain, distant rain streaks accumulate to form rain aggregates, the effect of which is similar to fog or haze and can scatter light, resulting in a significant reduction in scene contrast and visibility and significant degradation of image quality.

Single-image De-rainy algorithms have been applied to real-world life scenarios, such as target detection [1], video surveillance [2] and video tracking [3] vision tasks, which are based on the input image being clear and reliable, and therefore require De-rainy algorithms to De-rainy images with rain to ensure the effective operation of computer vision systems. The purpose of single De-rainy is to restore the rainy image with rainfall quality into a clean and rain-free image.

In 2012, single-image rain removal algorithms were first proposed [4], and early single-image rain removal algorithms mainly used sparse coding and dictionary learning to remove rain streaks from images; in recent years, with the development of deep learning theory and technology, image rain removal methods based on deep learning

H. Jin et al. (Eds.): GPC 2023, LNCS 14503, pp. 248–261, 2024.
https://doi.org/10.1007/978-981-99-9893-7_19

Fig. 1. The effect of rain on picture quality

have developed rapidly after 2017. These methods use deep networks to automatically extract rainy image features, and simulate complex image relationships from images with rain to clear images without rain by features. Single image De-rainy algorithms can be divided into two main categories [5], as follows:[1]

(*i*) Model-driven image-based rain removal algorithms. This class of methods uses a priori knowledge of the image, such as the direction, density and size of the rain streaks, to select a rain removal model, which is then solved by designing an optimization algorithm to obtain a clean and rain-free image. The main algorithms are image decomposition-based algorithms [4], discriminative sparse coding algorithms [6], joint two-layer optimization algorithms [7], and Gaussian mixture models [8]. Numerous improved algorithms based on deep learning have emerged on the basis of such algorithms.

(*ii*) Data-driven image-based rain removal algorithm. Based on deep learning, a nonlinear image from rain to no rain is learned by constructing a neural network using a synthetic rain streaks labeled image and a corresponding clean no rainy image. The main algorithms are JORDER [9], DetailNet [10], RESCAN [11], DID-MDN [12], PReNet [13], SEAN [14], DAD, SPANet, etc. Numerous improved algorithms based on deep learning have emerged on the basis of such algorithms.

In this paper, a single rainy image as the research object, residual neural network as the technical means, combined with the idea of recursion, rain streaks detection and removal. The main work is as follows: firstly, aiming at the introduction of the element attention mechanism, the features are weighted from the dimensions of space and channel, which improves the adaptability of the model to the rain removal task. Secondly the EAPRN model of synthetic rainy image and real rainy image and the method of recent years have obtained good results.

[1] This work was funded by the Foundation of Inner Mongolia Science and Technology Plan Funds(RZ2300000261), Natural Science Foundation of Inner Mongolia(2023LHMS01008),Doctoral Research Initiation Fund(DC2300001281).

2   Related Work

2.1   Model-Driven Image-Based Rain Removal Algorithms

In 2012, Kang et al. [4] first used bilateral filters to decompose the image into a low-frequency component containing the image structure and a high-frequency component containing rain and background texture, and then used dictionary learning and sparse coding to decompose the high-frequency component into a rain component and a no-rain component; the disadvantage of this method is that it often leads to over-smoothing of the image after De-rainy when the rain streaks is close to the background information structure; in the same year, Huang et al. [15] used SVM and PCA to classify dictionary atoms more accurately; in 2013, Kim et al. [16] observed that a typical rain streaks resembles an elongated ellipse, so the rain streak region was first detected by analyzing the rotation angle and aspect ratio of the elliptical core at each pixel position, and then the detected rain streaks region was filtered with a nonlocal mean value. In 2014, Chen et al. [17] introduced depth of field and feature color based on the literature [15] thus more accurately determining the rain-free layer and improving the rain streak removal effect, but the problem of rain streak residue or poor image background texture retention was common in the obtained rain removal images. In 2015, Luo [6] achieved single-image rain removal by discriminative sparse coding. The above sparse coding based De-rainy method has a wide range of application and can De-rainy both synthetic and real images, but the De-rainy effect is more general and more rain streaks remain in the image after De-rainy. in 2016, Li et al. [8] addressed the problems of too smooth background image or rain streaks residue in the dictionary learning, sparse coding methods and proposed to use two different Gaussian mixture models Gaussian mixture model (GMM) to model the rain streaks layer and the background layer separately, so as to separate and remove the rain streaks. This method is effective in removing lighter rain streaks, but is generally effective for heavier and sharper rain streaks, and still leads to the obliteration of background details to some extent. In 2017, Zhu et al. [7] constructed an iterative layer separation process to determine rainfall dominant regions using rain streaks orientation, which is used to guide the separation of rain streaks from background details based on rainfall dominant patch statistics. In 2018, Deng et al. [18] developed a global sparse model for rain removal, considering features such as the inherent directionality, structural knowledge and background information of rain streaks. In 2019, Xiao Jinsheng et al. [19] proposed to decompose the image using joint bilateral filtering and short-time Fourier transform, while introducing the concept of depth of field and using PCA and SVM to improve atomic classification accuracy, rain streak residual and for the background texture misremoval problem. In 2020, Son et al. [20] proposed a scribbled rainfall image set and a shrinkage-based sparse coding model that determines how much to reduce the sparse coding in the rainfall dictionary and keep the sparse coding in the non-rainfall dictionary, improving the rain streaks removal accuracy.

2.2   Data-Driven Image-Based Rain Removal Algorithm

In 2013, Eigen et al. [21] trained a special CNN by minimizing the mean squared difference between the predicted blocks of images with and without rain, and used deep learning for the first time to remove rain drops attached to images. 2017, Yang et al. [9] constructed a recurrent network for rain streaks detection and removal (joint rain detection and removal (JORDER), which was the first time that deep learning was used to remove rain streaks from a single rainy image; in the same year, to address the problem that it is difficult to simultaneously remove rain streaks and retain object features when the structure and orientation of the background objects are similar to the rain streaks, Fu et al. [10] obtained the relationship between images with and without rain by constructing a three-layer deep convolutional neural network (Detail Net) to remove the rain streaks. In 2018, Li et al. [11] proposed RESCAN (recurrent squeeze-and-excitation context aggregation net) to address the problems of insufficient utilization of spatial context information and failure to consider the intrinsic connection between the stages of rain removal based on the literature [9]. In the same year, Zhang et al. [12] proposed a density-aware multiplexed densely connected neural network algorithm (DID-MDN) to learn rain density labels and thus remove rain streaks. In the same year, Zhang et al. [12] proposed a density-aware multiplexed densely connected neural network algorithm (DID-MDN) to learn rain density labels and thus remove rain streaks, which is prone to blurring background details; meanwhile, Attention and Generative Adversarial Networks (GANs) have also been applied to the field of single-image rain removal, and Qian et al. [22] proposed AttGAN, which introduces a global attention mechanism in generative adversarial networks, and uses the generated binary mask image to guide the network for rain drop detection and removal. In 2019, Ren et al. [13] streamlined the network based on the literature [11] to improve the efficiency of rain removal time, designed a multi-stage PReNet, and designed an LSTM to aggregate contextual information between stages; in the same year, Zhang et al. [23] proposed a conditional generative adversarial network-based rain removal algorithm (ID-In 2020, Tan et al. [14] designed SEAN, introduced channel attention and spatial attention screening features, and proposed a joint rain streaks layer and background layer to jointly predict rain removal images. In 2021, Zhang et al. [24] proposed a rain removal algorithm based on channel attention and gated cyclic unit based rain removal algorithm (RMUN), which achieves parameter sharing in different loop stages through gated loop unit blocks in a recurrent network, thus better preserving texture and background information of the rain removal image.

3   Model Introduction

This chapter adds the elemental attention module to the PRN model to get the elemental attention-guided progressive rain removal network (EAPRN), The EAPRN model and the elemental attention module are described separately below.

3.1   EAPRN Model Framework

EAPRN uses an end-to-end design, where the input is a rainy image and the output is a rain-free image. The structure of EAPRN is shown in Fig. 2:

The EAPRN model contains three parts:

(i) distance judgment layer (E): Combined with Euclidean distance control model output.

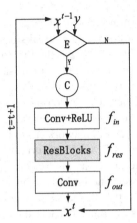

Fig. 2. EAPRN Model

(ii) image connection layer (C): Upconnection of the De-rainy image from the previous stage with the input rainy image
(iii) Residual Blocks module: The main network combining elemental attention mechanisms for predicting De-rainy image.

3.2 Network Architectures

The EAPRN networks all use a convolutional kernel of size 3 × 3 and a padding of 1 × 1. It consists of a convolutional layer and ReLU [22], containing 5 ResBlocks for one convolutional layer. The input is 6 channels and the output is 32 channels since it is connected by RGB three-channel input rainy image y and RGB three-channel stage result. The input and output of are 32 channels. The output of is used as the input and the output of RGB 3-channel De-rainy image.

Each ResBlock in f_{res} has 2 convolutional layers followed by a ReLU [22]. All convolutional layers accept 32 channels of feature information without upsampling or downsampling operations.

The elemental attention mechanism is used in the ResBlocks network, where each Resblocks module is followed by an elemental attention module, which reinforces the output features of each ResBlocks to get the elemental attention features, multiplies the elemental attention features with the output features to get the weighted output features, and then uses them as the input of the next ResBlocks. ResBlocks mainly play the role of rain streaks prediction and are the core of the EAPRN model. The introduction of elemental attention can divide the feature images extracted by ResBlocks into local regions and the importance degree between channels, and give higher weights to the features that are beneficial to the rain removal task, thus making the rain streaks prediction process more refined.

3.3 Element-Wise Attention Module

Elemental attention modules are mainly divided into spatial attention module $A_c(x_m)$ and channel attention module $A_s(x_m)$ belonging to $R_{H \times W}$ as shown in Fig. 3, where xm belonging to R_{HxW} are the elemental points of different channels of the input feature image. After the two attention modules are extended to the same size and dimension as x_m, they are multiplied to obtain the attention image $A(x_m)$ belonging to $R_{C \times H \times W}$. The element-level attention images are obtained by the spatial and channel attention modules, which are able to focus on all feature elements during the training process of the network.

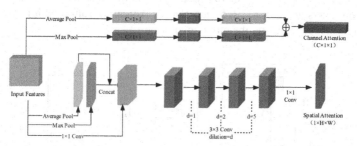

Fig. 3. Element-wise attention module

Channel attention focuses on the importance of features between different channels of the feature image as shown in Fig. 4, and the main purpose is to assign higher weights to the important feature images in the network. To reduce the computational complexity, global level pooling and global maximum pooling are used for each feature image to aggregate spatial information and generate two vectors $\{V_C, V_S\}$ belonging to $R_{C \times 1 \times 1}$. The two vectors are then fed to a shared fully connected (FC) layer and the outputs are summed to obtain the channel attention graph. The purpose of using the sub-method is to consider both the global and local information of the feature image.

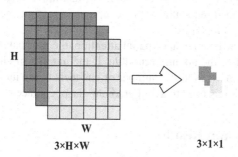

Fig. 4. Channel attention schematic

The number of simulated channels is 3, and the input features of height and width respectively and different colors represent the feature values of feature images on different channels. The Fig shows the average pooling to get the mean value of all the same color feature values, and similarly, the maximum pooling to get its maximum value. The above pooling operations are performed on the input features in turn until all channels are executed, i.e., the channel features are obtained.

The spatial attention pays attention to the importance of features between different locations of the feature images as shown in Fig. 5, and the main purpose is to assign higher weights to the important feature regions in the network. Similar to the channel attention module, first, the input feature images are pooled equally and maximally for elements at the same position in different channels to obtain two images $\{M_{avg}, M_{max}\}$ belonging to $R_{1 \times H \times W}$, which are input into three 3×3 hole convolutions and one 1×1 to finally obtain $A_x(x_m)$ belonging to $R_{H \times W}$. The dilation rate decibels of the three hole convolution layers are set to 1, 2, 5, and also, the attention values with large receptive fields can be calculated using the null convolution.

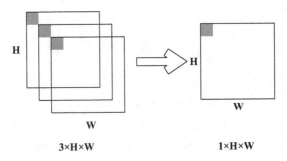

$3 \times H \times W$ $1 \times H \times W$

Fig. 5. Spatial attention schematic

The number of simulated channels is 3, and the height and width of the input features are summed respectively. The pink area shows the eigenvalues between the three channels in the input features at the $(0, 0)$ position. The Fig shows the average pooling to obtain the mean value of the feature values between each channel at the $(0, 0)$ position, and similarly, the maximum pooling is to obtain its maximum value. The above pooling operations are performed sequentially on the input features until the position is executed, which gives the spatial features.

After obtaining the channel and spatial attention, the residual connection is used for the input and output of the attention module. If the number of channels is different, a 3×3 convolution is used to fit the number of channels to the input feature image, and the residual connection is used to avoid gradient disappearance or gradient explosion.

4 Experiments and Results

EAPRN is retrained in Rain100H [8] (heavy rainfall dataset) containing 1800 rainfall images. In this paper, we use 4 DID-MDN [12], Rain12 [21], Rain100L [9] and Rain100H [8] four synthetic dataset datasets for testing.

4.1 Introduction to the Dataset

DID-MDN [7] contains 1200 rainfall images with different rain density, Rain12 [12] contains 12 rainfall images with different rain streaks directions, and Rain100L [9] and

Rain100H [9] contain 100 light rainy images and 100 heavy rainy images from different scenes, respectively.

Since Rain100H contains a large amount of rainfall line information as well as a large rainfall line density, which is very testing the rain removal ability of the model, the EDAMRN model is retrained in 1800 training set sample pairs of Rain100H (heavy rainfall dataset). It is also tested on 4 test set sample pairs of DID-MDN, Rain12, Rain100L and Rain100H. The experimental results are presented in four main parts: model loss function selection, recursive controller EDR threshold selection, synthetic rainy image results analysis, and real rainy image results analysis, which will be introduced separately below.

4.2 Loss Function

The model adaptively outputs De-rainy images at intermediate stages, so it is also important for each stage of supervised training, which is named the multi-stage loss function as shown in Eq. (1):

$$\mathcal{L}_2 = - \sum_{s=1}^{8} SSIM\left(x^s, x^{gt}\right) \tag{1}$$

Which x^{gt} is the no-rain plot, x^s is the De-rainy plot output from the s stage of the model.

4.3 Recursion Count

We select 100 pairs of images from Rain100H [30], put the rainy image into the network, and calculate the average values of SSIM, PSNR, MSE, and EDR between the De-rainy image and the clear rain-free image obtained from each iteration of all images as in Table 1.

Table 1. Quantitative assessment of the stage De-rainy image

Recursion	SSIM [27]	PSNR [26]	MSE	EDR [25]
1	0.9996	31.1748	49.6140	/
2	0.9969	31.3201	47.9809	/
3	0.9971	31.4687	46.3674	0.4795
4	0.9977	31.8622	42.3508	0.5185
5	0.9981	32.0691	40.3802	0.5973
6	**0.9983**	**32.1691**	39.4608	0.6750
7	0.9981	32.1594	38.9172	**0.7524**
8	0.9981	32.1565	38.9164	0.7962

We calculated the average results of each metric for the first 8 iterations of the network, and the definition of EDR dictates that EDR values cannot be calculated for the first two iterations. Since there is a large gap in the Euclidean distance between the first two iterations of the pictures, the number of iterations of the network must be more than two. As can be seen from Table 1, with the increase of EDR, the evaluation indexes SSIM, PSNR, and MSE are improved, but when EDR is equal to about 0.75, the three indexes show a decreasing trend.

Through the statistics in Table 1, in order to obtain better quality De-rainy images, we set the value of EDR as 0.75 as the condition to judge the termination of EAPRN iteration. The value of EDR is lower than 0.75, the network will continue to iterate downward, and when the value of EDR is higher than or equal to 0.75, the network stops iterating to output the previous stage of De-rainy images.

Considering some uncertainties, for some rainy image networks may lead to infinite loops, Fig. 4 we find that the Euclidean distance between two adjacent De-rainy images approaches zero after the eighth iteration of the network. The output De-rainy plot finds no visual difference between the eighth iteration De-rainy plot and the previous iteration De-rainy plot. Both of them indicate that this network reaches the upper limit of De-rainy capacity at 8 iterations. Therefore, we set the final number of iterations of the network to 8, which is used to force the network to stop constantly iterating.

4.4 Rain Streaks Chart with Attention Chart Display

As in Fig. 6, to facilitate the demonstration of the role of the attention mechanism, synthetic rainy images (a), (e) with different rain streaks densities in the same background are selected for testing in this section in Rain100H and Rain100L, and they are put into the predicted rain streaks images (b), (f) in the EAPRN model, before the de-rainy images (c), (g) are obtained by the additive composite formulation.

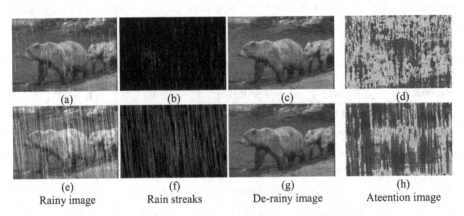

(a)	(b)	(c)	(d)

| (e) | (f) | (g) | (h) |
| Rainy image | Rain streaks | De-rainy image | Ateention image |

Fig. 6. Rain streaks and attention chart visualization

The rain streaks density in a is significantly lower than that in b. In the predicted rain streaks images (b), f, the rain streaks information is purer in (b), and the object edge information (blue) is obviously retained in (f). Then, by comparing the SSIM and PSNR of the De-rainy images (c), (g). It can be proved that the EAPRN model is more accurate in predicting rainy images with smaller rain streaks densities, and the quality of the De-rainy image is higher.

This section also visualizes the elemental attention in the EAPRN model for (a), (b). The vectors output from the elemental attention module in the model are shown by heat images, as shown in (d), (h). The redder the picture region represents the higher weight of the model for this region, and the (d), (h) corresponds to the rainy image (a), (e). It is found that the red region of the attention image corresponds to the rain streaks region of the rainy image, which is consistent with the original intention of the reference of the attention mechanism, and can help the model locate the rain streaks region and improve the adaptation ability of the model to the rain removal task.

4.5 Synthetic Rainy Image De-rainy Qualitative Analysis

The quantitative evaluation metrics for the rain removal method are peak signal-to-noise ratio (PSNR) and structural similarity (SSIM), respectively. The synthetic datasets tested were 100 test sample pairs in Rain100H, 100 test sample pairs in Rain100L, 12 test sample pairs in Rain12, and 400 test sample pairs in Rain12000.

Table 2 shows the mean values of the evaluation metrics obtained by the seven rain removal models on the four test sets, with the bold text indicating the best of these models and the numbers with horizontal streaks under them indicating the second best of these models.

Table 2. Seven methods tested on different datasets

Datastes	Rain100H		Rain100L		Rain12		Rain12000	
Method	PSNR	SSIM	PSNR	SSIM	PSNR	SSIM	PSNR	SSIM
RESCAN	26.45	0.846	33.11	0.9733	34.12	0.9615	29.88	0.9042
DIDMDN	25.00	0.754	30.87	0.9515	29.84	0.9326	27.95	0.9087
JORDER	23.45	0.749	32.46	0.9741	33.92	0.953	24.32	0.862
SPANet	29.55	0.9071	38.06	0.9806	35.34	0.9534	32.05	0.9442
DAD	30.85	**0.932**	38.01	0.9815	**36.72**	0.9533	33.67	0.9526
PReNet	29.46	0.899	37.48	0.979	36.66	**0.9617**	33.82	0.9654
EAPRN	**31.63**	0.915	**38.18**	**0.9834**	36.15	0.9582	**34.53**	**0.9717**

The EAPRN model proposed in the text achieves the optimal PSNR for Rain100H, SSIM and PSNR for Rain100L, and SSIM and PSNR for Rain12000. Compared with the suboptimal ones, the improvements are 0.78 dB, 0.12 dB, 0.0019, 0.71 dB, and 0.0063, respectively. The PSNR in the Rain12 dark scene dataset is lower than the optimal DAD model by 0.57 dB, and the SSIM is lower than the PReNet method by 0.0035. In summary, the EAPRN model proposed in this chapter has five optimal metrics compared with the six representative rain removal methods, it has 5 optimal and 2 suboptimal indicators, and becomes the optimal model among the above methods.

4.6 Qualitative Assessment of De-rainy Images

We select one real rainy image exhibit and one synthetic rainy image each, and output the De-rainy image in five models. To facilitate the comparison of the De-rainy plots, the red and yellow boxed areas in the Fig are zoomed in for the operation (Fig. 7).

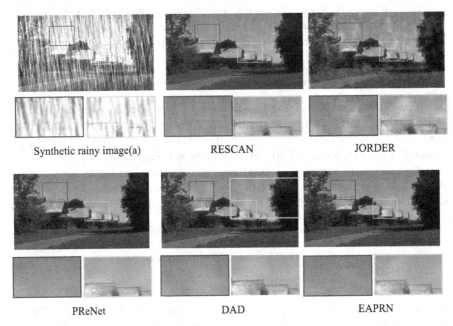

Fig. 7. Synthetic rainy image De-rainy comparison. (Color figure online)

For the synthetic rainy image (a), RESCAN and JORDER rain removal effect is similar to the synthetic rainy image (a), JORDER method the whole image due to rain streaks residue, the whole image whitish serious also exists block blur situation, RESCAN method the whole image is more blurred, the red box in the sky part can be seen streak-like rain streaks residue, EAPRN output rain removal image in the visual clarity has improved, DAD in the red box in the sky part of the whitish blur block excess PReNet, yellow box in the DAD, PReNet and EAPRN in the sky part of the local whitish phenomenon. The EAPRN model proposed in this chapter shows the least number of blurred blocks and the smallest area of blurred blocks (Fig. 8).

Observing the real rainy image (b), there is an obvious rain streaks residual phenomenon in JORDER, and the other four models have relatively good rain removal effect. The red box shows the recovery of two vertical buildings, and the local greening occurs in the images of PReNet, EAPRN and DAD, and EAPRN has the smallest greening area. In comparison, RESCAN and JORDER have better recovery of vertical

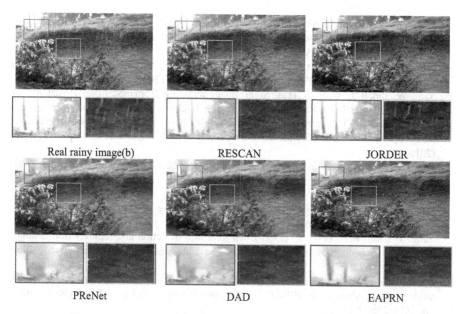

Fig. 8. Real Rainy image De-rainy comparison. (Color figure online)

buildings. However, it is found from the yellow box that rain streaks residue exists in both RESCAN and JORDER, and EAPRN has a more comprehensive removal of rain streaks and a higher purity of the image.

5 Conclusion

In this paper, we combine the elemental attention mechanism in the previously proposed PRN baseline rain removal model to design the elemental attention-guided progressive rain network (EAPRN), and put the elemental attention mechanism into the main Res-Blocks module of the network to obtain the weighted feature image, which improves the model's suitability for the rain removal task. The results of training on a large number of datasets and testing on commonly used synthetic and real datasets show that the effectiveness and rationality of the proposed algorithm are verified both in terms of subjective and objective evaluations.

References

1. Liu, S., Huang, D., Wang, Y.: Receptive field block net for accurate and fast object detection. In: Ferrari, V., Hebert, M., Sminchisescu, C., Weiss, Y. (eds.) ECCV 2018. LNCS, vol. 11215, pp. 404–419. Springer, Cham (2018). https://doi.org/10.1007/978-3-030-01252-6_24
2. Lee, K.H., Hwang, J.N., Okapal, G., et al.: Driving recorder based on-road pedestrian tracking using visual SLAM and constrained multiple-kernel. In: 17th International IEEE Conference on Intelligent Transportation Systems (ITSC), pp. 2629–2635. IEEE (2014)

3. Forero, A., Calderon, F.: Vehicle and pedestrian video-tracking with classification based on deep convolutional neural networks. In: 2019 XXII Symposium on Image, Signal Processing and Artificial Vision (STSIVA), pp. 1–5. IEEE (2019)
4. Kang, L.W., Lin, C.W., Fu, Y.H.: Automatic single-image-based rain streaks removal via image decomposition. IEEE Trans. Image Process. 21(4), 1742–1755 (2011)
5. Yang, W., Tan, R.T., Wang, S., et al.: Single image deraining: from model-based to data-driven and beyond. IEEE Trans. Pattern Anal. Mach. Intell. 43(11), 4059–4077 (2020)
6. Luo, Y., Xu, Y., Ji, H.: Removing rain from a single image via discriminative sparse coding. In: Proceedings of the IEEE International Conference on Computer Vision, pp. 3397–3405 (2015)
7. Zhu, L., Fu, C.W., Lischinski, D., et al.: Joint bi-layer optimization for single-image rain streak removal. In: Proceedings of the IEEE International Conference on Computer Vision, pp. 2526–2534 (2017)
8. Li, Y., Tan, R.T., Guo, X., et al.: Rain streak removal using layer priors. In: Proceedings of the IEEE Conference on Computer Vision and Pattern Recognition, pp. 2736–2744 (2016)
9. Yang, W., Tan, R.T., Feng, J., et al.: Deep joint rain detection and removal from a single image. In: Proceedings of the IEEE Conference on Computer Vision and Pattern Recognition, pp. 1357–1366 (2017)
10. Fu, X., Huang, J., Zeng, D., et al.: Removing rain from single images via a deep detail network. In: Proceedings of the IEEE Conference on Computer Vision and Pattern Recognition, pp. 3855–3863 (2017)
11. Li, X., Wu, J., Lin, Z., et al.: Recurrent squeeze-and-excitation context aggregation net for single image deraining. In: Proceedings of the European Conference on Computer Vision (ECCV), pp. 254–269 (2018)
12. Zhang, H., Patel, V.M.: Density-aware single image Deraining using a multi-stream dense network. In: Proceedings of the IEEE Conference on Computer Vision and Pattern Recognition, pp. 695–704 (2018)
13. Ren, D., Zuo, W., Hu, Q., et al.: Progressive image deraining networks: a better and simpler baseline. In: Proceedings of the IEEE/CVF Conference on Computer Vision and Pattern Recognition, pp. 3937–3946 (2019)
14. Tan, Y., Wen, Q., Qin, J., et al.: Coupled rain streak and background estimation via separable element-wise attention. IEEE Access 8, 16627–16636 (2020)
15. Huang, D.A., Kang, L.W., Yang, M.C., et al.: Context-aware single image rain removal. In: 2012 IEEE International Conference on Multimedia and Expo, pp. 164–169. IEEE (2012)
16. Kim, J.H., Lee, C., Sim, J.Y., et al.: Single-image deraining using an adaptive nonlocal means filter. In: 2013 IEEE International Conference on Image Processing, pp. 914–917. IEEE (2013)
17. Chen, D.Y., Chen, C.C., Kang, L.W.: Visual depth guided color image rain streaks removal using sparse coding. IEEE Trans. Circuits Syst. Video Technol. 24(8), 1430–1455 (2014)
18. Deng, L.J., Huang, T.Z., Zhao, X.L., et al.: A directional global sparse model for single image rain removal. Appl. Math. Model. 59, 662–679 (2018)
19. Xiao, J., Wang, W., Zou, W., et al.: An image rain removal algorithm via depth of field and sparse coding. Chin. J. Comput. 42(9), 2024–2034 (2019)
20. Son, C.H., Zhang, X.P.: Rain detection and removal via shrinkage-based sparse coding and learned rain dictionary. J. Imaging Sci. Technol. 64(3), 30501-1–30501-17 (2020)
21. Eigen, D., Krishnan, D., Fergus, R.: Restoring an image taken through a window covered with dirt or rain. In: Proceedings of the IEEE International Conference on Computer Vision, pp. 633–640 (2013)
22. Nair, V., Hinton, G.E.: Rectified linear units improve restricted Boltzmann machines. In: ICML (2010)
23. Zhang, H., Sindagi, V., Patel, V.M.: Image De-rainying using a conditional generative adversarial network. IEEE Trans. Circuits Syst. Video Technol. 30(11), 3943–3956 (2019)

24. Yan, Z., Juan, Z., Zhijun, F.: Image rain removal algorithm based on channel attention and gated recurrent unit. Appl. Res. Comput. **38**(8), 2505–2509 (2021)
25. Chen, X., Ma, R., Dong, Z.: Research on single image deraining algorithm based on euclidean distance with adaptive progressive residual network. In: 2021 IEEE 23rd International Conference on High Performance Computing & Communications; 7th International Conference on Data Science & Systems; 19th International Conference on Smart City; 7th International Conference on Dependability in Sensor, Cloud & Big Data Systems & Application (HPCC/DSS/SmartCity/DependSys), pp. 1791–1796. IEEE (2021)
26. Hore, A., Ziou, D.: Image quality metrics: PSNR vs. SSIM. In: 2010 20th International Conference on Pattern Recognition, pp. 2366–2369. IEEE (2010)
27. Wang, Z., Bovik, A.C., Sheikh, H.R., et al.: Image quality assessment: from error visibility to structural similarity. IEEE Trans. Image Process. **13**(4), 600–612 (2004)

Author Index

Printed in the United States
by Baker & Taylor Publisher Services

Printed in the United Statesby Baker & Taylor Publisher Services